A PEBBLE in a POND

To Jordan & Susan,

I am so glad we met... Enjoy!

Sincerely,

Layne T. [illegible]

3 Aug 2016

A PEBBLE *in a* POND

LAYNE T. OLIVER

LIFE IS AN ADVENTURE THE JOURNEY IS FULL OF DRAMA - GOOD AND BAD AND LIKE THE WORDS WE SAY...THE ACTIONS WE TAKE AND THE DECISIONS WE MAKE, CREATE RIPPLES... JUST LIKE THROWING-

TATE PUBLISHING
AND ENTERPRISES, LLC

Published by Tate Publishing & Enterprises, LLC
127 E. Trade Center Terrace | Mustang, Oklahoma 73064 USA
1.888.361.9473 | www.tatepublishing.com

Tate Publishing is committed to excellence in the publishing industry. The company reflects the philosophy established by the founders, based on Psalm 68:11,
"The Lord gave the word and great was the company of those who published it."

Cover design by Samson Lim
Interior design by Mary Jean Archival

Published in the United States of America

ISBN: 978-1-68164-474-5
Biography & Autobiography / General
15.06.30

After a decision is made, there is potentially a second- and third-order effect, maybe even more.

Layne Oliver, a.k.a. Lightning Layne, has spent a lifetime experiencing and learning about these effects. By the time he made it to eighteen years of age, he was lost and aimless; he barely made it out of high school and really saw no future for himself.

However, a decision to move back in with a parent who exhibited symptoms of what now may be called IED (intermittent explosive disorder) put him on a pathway where he would meet numerous positive mentors. These individuals would instill (grind in) the discipline and core values necessary for this young man to face and handle the many challenges life would bring his way—even his own imminent death. The lesson here is there are a multitude of people with problems that make yours seem trivial!

You can survive anything if you want to, but you have to have a reason to live that hurts more than the hurt that wants you to give up. How you choose to survive or not becomes the final question.

To the guys who showed me the way, my brothers—Dennis and Jess, Jeff Lowe, Wayne Gregory, Harry Allred, and power-quality specialist John Jackman.

To my sisters—Linda, Tanya, and Elaine—thanks for all the love.

I also want to recognize the many friends of my childhood, especially David Viall, Mark Gomez, Matt Fife, Melvin Healy, Ivan Udy, and Chris Beck. You guys filled in the holes that my family couldn't. We had a great time in Boy Scouts, school, and life; a recent class reunion reminded me of this.

But most especially, I want to recognize the warm and tender woman I love, my wife, Teri. For without her, I would be nothing and aimless! To our children: Angela, Melanie, Eric, and Amy—and their "mates": Taleb, Tommy, Madelynn, and Juan V., including our grandchildren: Simon, Andrew, Bridger, Brielle, and little Gunner, who have all brought so much light and love into our world.

Acknowledgements

Favorite Authors

Ralph Young, Ted Kerasote, Bruce Lee, Chuck Norris, John Long, David Roberts, David Levy, Napoleon Hill and Sir Chris Bonnington, Og Mandino, James Edward Mills, and Edward Abbey.

You helped me dream and open my mind and brought me calm!

Climbers

Don Larson, Richard Wright, Jay Night, Fred Dirtschi, Steve Garner, Eric Hill, Keith Phelps, Bobby Sorenson, Steve the Electrician, Raleigh Collins, Jeff Lowe, Merrill Bitter, Doug Heinrich, Conrad Anker, Seth Shaw, Tony Calderone, John Bacher, Tim, Scott, Paul, Andy, Jason and Colby, Ray Dahl, Dave Smith, and everyone else.

Whether I shared a rope with you or you sold me gear to help get me out there, you guys are amazing and you helped me dance. Thanks.

Karate Instructors and Heroes

Senseis Toshio Osaka and Rick Hobusch, Bruce Lee, Chuck Norris, and Bill Wallace.

You guys taught me how to face life!

Active Duty and National Guard Buddies

Thanks for standing tall in the gap and protecting freedom and being flexible with me!

The world is a better and safer place because of your heart, your skill, and your diligence!

Field Testing

A special thanks to all the companies who trusted me to test their equipment for outdoor pursuits, especially Wayne Gregory of Gregory Mountain Products for getting me started. Also, Russ Pittman of Terramar Sports, Jenny Dow and the crew at Fox River, the Asolo boys, Marco and Luca Zenatta, Bruce, Stu, and Jason, Big Bill and Little Bill of Swaney Gloves, and Sally McCoy and Fred Dieter of Sierras Designs.

All of you have great faith!

Finally

Cheers to all the folks who read my manuscript and encouraged me and offered advice—beginning with Wayne Gregory, Russ Pittman, Mark Zodda, Kathy Bates, Robert Cole, Dr. Daniel Guthrie, Marlin Stum, Stephen Johansen, Joyce Hayes, Jess Lowe, and especially Amy Oliver, Melanie Speirs, Eric Oliver, and my wife Teri.

I want to extend a special thanks to Tate Publishing for wading through all my wanderings and guiding me through the publishing process, nudging me, encouraging me, and teaching me. Without the help of the staff at Tate, this book would still be a compilation of work needing lots of help.

Where would I be without my family and friends?

Life is an adventure
The journey is full of drama—good and bad!
And just like the words we say,
The actions we take and the decisions we make create ripples…
Just like throwing
A pebble in a pond.

Contents

Author's Note

The stories in this book are based on actual experiences that I was involved in or that were related to me from other participants from a specific event; if there are any inaccuracies, the fault is mine. I have, out of common courtesy, changed some of the names of people, places, etc. to protect the anonymity and confidentiality of the persons involved. In other cases, I have told the story exactly as it was, including referencing the actual players involved.

Introduction

I started life like many do. I was born "breech first" to a mom and a dad—Marjorie and Bill Oliver. It was September 18, 1955, at St. Marks Hospital in Salt Lake City, Utah.

My older sister Linda and two older brothers, Dennis and Jesse, were already being taught the authoritarian process by our dad, who was their stepfather. Now I would learn this process as would eventually my two younger sisters and our many step- and half siblings as life evolved. My mom left dad when I was five; my siblings and I remained fatherless from then on, which I am still sore about! The abuse of my childhood instilled many things in me, particularly an independence and a need to succeed that drives me mad and hard to this day, even though sometimes my inner feelings make me want to just give up too.

By some fortune, something deeper keeps me hanging on; things do change, and life seems to balance things out as I have been lucky to have had my mom, brothers, sisters, and friends along the way to support me and help fill in the many voids. Through my older brothers, Dennis and Jess, I began my "woodsmanship" training when I was around seven years old. I learned to hunt and fish, cook over a fire, and camp. I learned how to safely handle firearms and shoot pretty straight. They led me in a direction that would cause me to become several things, for example: a mountaineer, astronomer, writer/reader, cyclist, field tester, photographer, public speaker, musician, and a model car and airplane enthusiast.

Despite it all, my dad helped me become a military man, which led to my becoming a commercial electrician! Life can take you in so many directions, and it can be exciting as well as confusing. I found myself content and successful in each of these endeavors.

I stand amazed because my childhood instilled such deep fear and anger in me to include the feeling of being second class and sometimes even worthless; it's so hard to get anywhere with those sort of misgivings about oneself! Eventually, I pushed myself to overcome every goal and obstacle in front of me, and along the way I often missed the forest for the trees too.

During my thirties, I was preparing myself to climb really big mountains and was constantly out there building my mental toughness and skills when I came face-to-face with something that had the power to end my life, and it would become a huge lesson in life and death. Because of this, I eventually realized that no matter what I was going through, that my pain and other hurdles were insignificant when compared to other folks' problems, especially in light of the greater good. I simply had to be strong and adjust with my circumstances. Was it, or is it, easy? No.

These days, I'm a husband, a father of four, and a grandfather of five—who all know me as "Pop," and I'm known for being there for all of them. I have to be strong for them and be around to help guide them! I had to be strong for my Air National Guard Unit too in which I served the country for thirty-two years. The National Guard was also my civilian workplace for about twenty-four years, and there were lots of challenges associated with this part of my life, and it may well be that I will not have learned everything I needed to from this collective experience before I leave this earth.

I also have to be strong for the people I teach the buddy system and mountain safety to (i.e., the Boy Scouts). I have something important to add to their skills and knowledge, and if I had given up, my voice and ideas would not be heard and this action just might cause someone to have it a bit harder than they should. I needed to be strong for my climbing partners too as this relationship is very much like a marriage where total trust is a minimum requirement and you simply become united with these folks. All these people have been so strong for me too; we simply love and support one another to the end!

Finally, I have to be strong for me. My life has been one learning experience after another, and I seemed to have grown to the understanding that there is something positive to be gained out of even the worst situation: that you can get through it if you really want to, and it offers you something valuable even if you don't recognize it right away! Through all of this, I've come to realize that the decisions we make and the actions we take, even the words that fall off our tongues, create effects that go on and on, forever and ever. Just like the ripples from a pebble thrown into a pond.

The Front Door

1

Life Begins

Life for me, in many ways, began in 1975 when I joined the Air National Guard, but before that time however, I had enjoyed many things like racing slot cars, flying model airplanes and rockets, which I had—for the most part—bought working odd jobs. I had fished and had hunted rabbits and deer, did lots of camping, played countless games of Ping-Pong with my brother Jess and played basketball with my friends, and I was a Boy Scout.

All of these sports and hobbies were not only fun but educational too. My two older brothers Dennis and Jess had introduced me and several of my friends to many of these pursuits, which helped us focus on something positive instead of drugs, alcohol, and chicks. By the time we were in junior high, we already knew that if you kissed a girl, she would get pregnant, so they were out of the picture for now.

Junior high through high school had been difficult. I worked nights at a grocery store to help pay for my and my sister's school tuition, which helped my mom make ends meet.

My mother had applied for Welfare to help us survive; it gave us food and some clothing, but it was for me so embarrassing, especially the free lunch program at school and the food stamps. When the other kids found out you were poor, some made a lot of fun of you! I found out, however, that a roof over your head and food to eat are better than the alternatives.

My older siblings were now out of the house and on their own at this point in time. One of them, Jess, had joined the Air Force when he was seventeen to get out of the way. At times, things were so tight around the house moneywise, that he would put water in his cereal so the three younger kids could have milk in theirs.

Boot camp for Jess was hard; it was the midsixties, and DIs (drill instructors) were allowed to pound you if they felt like it! Nonetheless, my brother's sacrifice didn't go unnoticed; little kids recognize much more than they are given credit for! Our dad never sent child support, took care of us at birthdays or Christmases, etc.; he never made an effort to let us know we meant more to him than being convenient punching bags. I always wondered if our dad ever cared about what his flesh and blood aspired to.

When we were younger, my oldest brother Dennis would buy our Christmas gifts and help raise us, and I know it was hard for him, but he loved and cared for the rest of us very much. As we grew up, bad habits and associations were around every corner just waiting, always waiting to deliver their cruel and crushing blows. My mom kept vigil on her children and protected us and guided us along. She routinely read to us from the Bible and other religious books and took us to church regularly! One thing I will never forget about her was despite our meager allotments for food and money, she did some baking and made some of the best sugar cookies around! I know she worried herself to sleep routinely, what a life!

Isn't it amazing how one individual caused us to run away out of fear? Someone that should have cared for us and loved us; after all, he helped bring several of us into this world. We were forced to live a life that was much harder and bleaker in many ways than it normally would have been. We left that prison never to return and gained some version of freedom even though it took a while for me to understand this.

Momentarily, I will talk more about why we had to run away.

A decade and a half later, right after I graduated high school, for whatever reason, my dad began communicating with me on a regular basis. He lived in Boise, Idaho, with my stepmom, Laretta, and her two boys, Lance and Brett! I to this day consider these people as family. I love them and am very proud of all they have achieved. Laretta is so much like my own mother; she really understands life and has lived through her share of hard times, even the Great Depression. My dad thought it would be a good idea if I left Kearns, a suburb of Salt Lake City, and move in with him and Laretta, insisting that there are two sides to every story. He said those words knowing that my mom would think it would be a bad idea and caution me against going through with it. I think my dad had forgotten that when I was five years old, my brother Jess, my sister Tanya, and I watched as he beat up our mother; this happened many times! He had also beaten up my brothers and sisters and me too! We saw it all. We lived in Blackfoot, Idaho, back then; and this first five years of my life were full of brutal attacks to everyone in our family. As a routine, he would make my brother Jess take risks that seemed to be for his personal enjoyment. One time he made Jess slide from the top of a spillway at a reservoir dam on a car inner tube, down to the river it joined; it was a very long way to the bottom. When Jess, traveling at breakneck speed, slammed into the river, he was immediately caught in and got tangled and tore up by barbed wire that was lying at this junction in the water!

I remember my brother Jess getting in the car shaking with puncture wounds all over him and our dad laughing about the incredible event; Jess looked on and looked scared! Another time, we were picking Jess up from a potato farm where he worked picking "Idaho potatoes"! He had ridden his bike to work that day, and we needed to go get him for whatever reason. When we arrived at the potato farm, our dad suggested that before we put Jesse's bike in the trunk that we should tie it to the car with a long piece of wire and have Jess ride the bike while it was being towed behind the car. I remember the look on Jesse's thirteen-year-old face as we sped down that bumpy country road; he looked terrified, and the bike was skittering side to side! We were doing thirty to forty miles an hour when the wire broke and my poor bother went careening off the road into the sagebrush, flipping the bike over and over and sending Jess flipping through the air then piling up on the sagebrush and dirt. Luckily (I guess), Jesse's only physical injuries were cuts, scratches, and bruises, but the bike he worked so hard for was ruined! When we got home and my mom found out what had happened, she threatened my dad, which of course turned into more beatings for the lower echelon; he seemed to always win. Around this same time, it became my turn in the barrel.

One day, some kids I knew were riding their bikes on a road behind our backyard and said something to me that caused me to shower them with pea gravel! My dad, hearing the commotion, saw me throw some more gravel; yelling out our back door, he ordered me to come in the house right now. So I ran around the house and came in the front door, I noticed that he was sitting in a chair at the kitchen table. When he saw me come in, he said in a very scary gruff commanding voice, "Come here." I slowly approached him scared to death! I remember my little hands being clasped together in front of me and being subdued with fear; suddenly his big fist was coming at my face. He had broken my nose and knocked me unconscious, this damage became more damage for my mother as she fought back to defend her child from a respected volunteer policeman, military man, umpire, and businessman! She was beat up one time because he slugged me in the kidneys, causing me to have blood in my urine.

There were times when things were going to become very violent in our home, so our mother would find neighbors who were willing to hide us kids away until the storm was past! I remember one event very well. A war was taking place in our home and we kids had been hidden in this bald guy's house. He had us holed up in a closet and would bring us food and drink to be consumed inside this closet space; I can still see his be-speckled face handing us food through the partially opened door. The funny thing is I remember only myself and my two little sisters hiding out! I asked Jess what he remembered, and he said that it was us four in there, and as for the bald guy, he was his junior high principal.

My family had become victims and prisoners of a tyrant, afraid of shadows and even the light, eventually, just plain afraid. There were so many days full of anguish and fear that they seem to run together like a torrent. Looking back, I wonder how it could have been allowed to happen; but our minds have buried the reality deep within adding insult to injury. It was an awful way to exist! Remember too that I have an older sister and brother who went through similar challenges that eventually drove them away; being away, in a way, they were lucky, but I wonder about their deepest feelings? Not everyone can or will express those hurtful feelings that swim around in their minds for a lifetime!

One day as I was coming home from West Bridge Elementary School, where I was a first grader, I noticed a white truck and a U-Haul trailer in our driveway. As I entered the house, I was shocked at the scene before me! Our front room was destroyed; all the furniture had been thrown down the stairs leading to the basement and was broken to pieces. The walls were full of big gaping holes, just smashed in! It looked as if a car had been in there backing into everything. The table lamps, vases, etc. were in pieces spread all over—in that place there had been another war! Apparently the rest of the house had also suffered similar damage as well. Presently, my mother came from another room to see who had come into the house, fearing that it might be her husband. She grabbed me and hustled me out to the waiting truck and told me to stay there. My mother wanted to avoid any kind of confrontation with her husband, but Dennis, my oldest brother, was prepared for such an event! When Dennis left home several years back, he was a member of the Utah Air National Guard and had purposely taken every self-defense course that the Air Force offered to help level the playing field with his stepfather if ever need be. Our mother had planned an escape and had called my oldest brother to come and get us; so Dennis showed up at a predetermined time and helped us make our getaway back to Salt Lake City, which is where we are all originally from.

2

Escape

We made it back to Utah and moved in with Mom's younger sister Aunt Niece and Uncle Frank and my nine cousins (soon to become twelve) in a suburb of Salt Lake City known as Sandy, which back then was considered the country. Kind of a nice place to live as it was calm and quiet. I am sure that our presence put a strain on everyone, but I never heard about it ever! That family was always willing to help even if they really couldn't; I remember our stay as being a good time, which was a definite change from how things had been. I was always tense and a little shell shocked by the different lifestyle. My aunt Niece liked to call me Tiger, which made me feel special, and she treated all of us with great love and kindness! Several months later, my mom and oldest bother, Dennis, found us a duplex that we could share right in Salt Lake City. Dennis had suffered a divorce, so pooling our limited resources seemed like a good idea, plus we were only a short drive from Aunt Niece's place, which kept the family close together if anyone needed help. My mother applied for

State Welfare and LDS Church welfare programs to help make ends meet. (Note: The Latter-day Saints church has programs that help supply food and clothing to people in need and offers volunteer work and paying type jobs as well.)

And so we started a new life in Salt Lake City, a little humbling and rugged, but we were no longer prisoners in someone else's version of life! My sister Tanya and I were enrolled in school at Washington Elementary, a block east of our duplex, which we rented from a man whom we called Grandpa Bement. The church we would go to was right next to the school, so we were set. Our youngest sister, Elaine, would eventually start going to kindergarten soon, but for now, she would stay with our mother, who managed our new home. The Bements, who lived several doors down the street, became good friends with our family and us kids played constantly with their kids, ate dinner at their house, went to little league ball games, Fourth of July fireworks at Liberty Park nearby, etc. Kenny and David Bement and I were generally the same age; we flew model airplanes together and were on the same little league baseball team that was known as Christiansen Diamonds! Local companies would sponsor team sports back in the day, and the team would adopt the company's name, which would appear on the uniform!

Time was passing relatively smoothly, and after a couple of years, we were just a natural part of the neighborhood. One sunny day while at school during recess (I was seven years old), I was outside running the so-called track, which was really just the perimeter of the asphalt school ground with a painted line representing the path. When according to my classmates, a large man whom nobody recognized came onto the school grounds! I never saw this person who came up behind me as I was running along and hit me with a board. When he was done, I was unconscious and he slipped away. Some of my classmates ran for help while several others carried me to the office. The only thing at all that I remember was that I was running the track and was wearing a blue and white Western style shirt with the little snap down breast pockets and then the ground was coming at me; at that moment, the lights went out, and then I woke up in a hospital! I had been taken to the Salt Lake County Hospital in South Salt Lake and had been diagnosed with a concussion, contusions, and had several broken vertebra! I wasn't able to walk very well for several months, and I remember being made to sleep on a very hard flat surface for several months. It's worth noting that in those days, my father regularly visited a sports shop in Salt Lake that he was part owner of. We never saw him when he came to town nor did I see my attacker. Twenty years later, while at work, I suddenly stopped what I was doing! My inner voice told me who the attacker was; I was thinking it was my dad! Internally, there was no question that it was him; he wasn't done hurting our family.

To this day, I struggle because of this injury and have suffered through long doses of traction that have you gripped in constant pain! However, the end result is usually positive, muscle spasms that can bend you in half get repaired and you can look to be an inch taller. This particular injury, because of the spasms, has made life difficult; many things can cause a spasm to occur or my vertebrae to go out of alignment (e.g., carrying heavy rucksacks (military slang for a comfy backpack). Also being an electrician, you can get into some very tight spaces that contort your body and bring on these painful and debilitating so-called charley horses and alignment issues. I have found that having good muscle tone and a good stretching routine have all but eliminated the problem, although at times I will need a doctor to help with medicine or some form of physical therapy as necessary. In my mind, very few things in life are insurmountable, though a lot of work may be necessary! I believe you can almost always find a reasonable solution to a challenge if you want to; it simply is a matter of perspective and choice. I know I'm making overcoming obstacles sound easy, but the truth is, overcoming a problem might mean that you have to learn to live with it and manage it because it has moved in for the duration. My blind faith is this: the darkness will fall, the next day will come—the moon will rise as will the stars and the sun—they will, such is life!

One morning while recovering from the beating, my grandpa Wally came to my bedside to comfort me and talk. I noticed that Grandpa Wally had his coin purse in one of his hands. Grandpa Wally always gave us kids quarters out of his old-fashioned coin purse, which was oval shaped; you would squeeze the sides causing a slit through the center to open. True to his habit, he started off by handing me several quarters. When we would visit him and Grandma, he would play his stereo real loud, insisting that unless music is played loud, you can't hear the full stereo effect! He would also dish out the vanilla ice cream cones to anyone who wanted or needed one! Everyone loved him very much. He was a great man, wise and tough, soft-spoken and full of love, and he could really catch fish. Alas, he was a man of God! His faith was unmovable and he displayed great pride in his posterity! While he was visiting with me, he gave me a hat pin; it was a rifle and was something he liked very much! Even as a seven-year-old, I actually very much appreciated the value and meaning of this gift. Grandpa Wally then told me that God had plans for me, a special mission; prompting me to stay strong, he said to let God guide me throughout my life and that all things happen for reasons unknown. Finally, he encouraged me try to make the best out of every situation! This is very good advice for everyone's sake! He told me to be timely, to be a man of my word, undeniably honest in all my dealings, and to help others; it was pretty much the Ten Commandments and the Boy Scout oath rolled into one! He also asked me to quit playing with matches as me and one of my cousins, Kim, almost burned down his garage by just flicking lit matches indiscriminately into weeds. Fortunately a neighbor was paying attention to the two little "pyros" and dowsed the fire with his garden hose before it could get out of control. But a fireman still gave us the "wrath of God" speech, scaring us silly by telling us that we now had a record! In the end, I promised Grandpa Wally no more playing with matches!

After fifty years of morphing in the boiling pot of life, I realize now just how much wisdom that old grandpa of mine was shoveling my way. Interestingly, Grandpa Wally's words about a mission came true—and in more ways than one. Also, there was a significant event coming that would change my life forever, and everything that I had experienced before its time was in a way preparatory. Around this time, my mother had met and married a man named Bud. I really don't remember too much about Bud, except that he took us camping often and he and my mother fought a lot (minus the hitting). Bud also drank a lot and he had a glass eye, which seemed to always stare with a serious tone, straight ahead! There were times when it was hard to tell which eye was the glass one! When we went camping, not owning tents or sleeping bags, we would fashion pup tents out of blankets stretched over a rope tied between trees, not fancy by any means, but it worked okay. We would then wrap up in blankets side by side and shiver the whole night! Sometimes I would opt to sleep in the trunk of the car, but I was afraid of the Mummy and King Kong. I had seen them on TV tearing things up and was sure that one of the two would get me during the night, so I often wound up back in the tent. Bud always made sure that we brought along a big bucket of Kentucky Fried Chicken, at least for the first dinner. With our makeshift tent, the chicken, the sound of a nearby stream, and the fresh crisp mountain air, everyone was smiling and seemed to be very happy, life was pretty good! I remember those trips as being fun; the kids and the adults played hide-and-seek and a fun card game called Fish. We even caught and ate a few fish! We would cook hot dogs over a fire and bake potatoes and trout in the coals; we would cook other things like eggs, bacon, and chili on an old Coleman white gas stove. We always had a campfire at the end of the day and roasted marshmallows, told scary stories till bedtime, and the next day, there was always a nice warm fire to wake up to! Something I have found to be quite interesting is the power of association (i.e., the smell of Coleman fuel brings back the memories of those camping trips so very long ago!). I have also experienced this on many recent mountaineering and/or backpacking trips where white gas was being used in a mountain stove. Those memories are sweet!

I mentioned before that Bud had a glass eye! As the story goes, he had lost the original eye years ago while on a ship deck in the Navy. Apparently, a big crane arm was moving his way when its hook caught on something then released, swinging very fast, hitting him in the right eye! Bud seemed to misplace his glass eye on occasion and would be squinting and searching all over for this interesting fitting. One time, one of my cousins, Billy (my mom's brother's (Bill's) son), was staying over for the weekend. Well, Billy could be a character at times and he had taken Bud's eye and hid it, so there was Bud looking all over the house asking if anyone had seen his eye. Billy, in his smart-aleck way, asked, "Ain't it in your head?" We kids all laughed; however, Bud (Mr. Serious) seemed less than amused, and all the more when he suddenly noticed that in our fishbowl was a big eye lying on the gravel-covered bottom, watching the gold fish that seemed to be curiously watching it.

About a year later, we were all down at Utah Lake near Provo Utah Boating with my uncle Bill, his wife, Aunt Darlene, and their kids. My uncle Bill believed the more the merrier and enjoyed helping others have a good time and would literally give you the shirt off his back and would go hungry so you wouldn't have to!

My cousin Larry who was six years old at the time (Uncle Bill's youngest son) and I were fishing for catfish together off the end of boat dock (we were fishin' pals); already, some stinkin' catfish had pulled one of our rods off the dock and into the water. Presently, my brother Dennis was standing directly behind me teaching us some fishing lesson about watching your fishing rod when one of our girl cousins, Val, came up from behind to push him into the lake! She hadn't realized I was right in front of Dennis, so when she gave him the shove, all I remember is the lake coming at me, then the lights went out! After a few moments (I suppose), I could see bubbles rising away from me toward a brightness—everything else around me was darkness. I could see my arms and legs above me as I calmly sank into the lake! Bud had seen me and Dennis go into the water and had scrambled to come and help! Dennis (I believe) was the one to pull me from the water; he said it was so murky that as he dove in deeper, he scooped his arms around and luckily found me and got me to the surface where Bud and several others could pull me up onto the dock. I woke up on the dock having first aid administered to me. Water from then on had become a big fear of mine!

Eventually, I took swimming lessons and learned how to swim and be more confident in water, which is good because I wound up getting into canoeing, which involves lakes and rivers, which also are places where one can drown.

Fear is a huge obstruction in life, but one can overcome it if one wants to! It will require knowledge and experience and cheek biting and faith though. Facing your fears is very difficult, but to overcome them, it is what you must do! Sometimes all you need to go on is nothing more than a little hope and blind faith. Again, if I'm making things seem easy, I apologize, but there is always a solution, always!

Anyway, the drug and crime scene was growing fast in Salt Lake City right along with the heat of the Vietnam War! There was a lot of bad influence where we lived, so my mom decided to move us all out to a little suburb west of Salt Lake City known as Kearns! Kearns at that time was considered to be something like the West Side Story; if you lived there, you were considered to be second class. I noticed this air right away, but as I aged, I realized that it wasn't true at all. Lots of good people lived there, people who cared, people who served others to include the community and their country, and a few of them even died for their country, the United States of America! Sure, we had our problems, but everyone does. Glasshouses suck!

3

Kearns

I began fifth grade in 1966 at my and my sisters' (Tanya and Elaine) new elementary school—South Kearns Elementary; my new teacher was Mr. Clarke. We rented a home from a county sheriff named Dale Gates; the house was located at the northeast corner of the school grounds.

By now, at home, it was just Tanya, Elaine, our mother, and me; Jess was still serving in the Air Force and had gotten married to a Cheyenne girl named Kathy! Dennis was again out on his own, working at a gas station in the daytime and playing guitar in a band at night. Our older sister Linda was married and raising a family and wading through her own tremendous struggles that would go on for the rest of her life. We survived on welfare provided by the State of Utah and the LDS Church, which we were members of. I worked on occasion at the church farms and canneries, just like my brother Jess had done, to help pay for our allotments. It was the only way we could look at ourselves in a mirror and feel a little bit of pride. School for me went okay but, I just couldn't seem to focus and keep up, so my grades were not very good. Our mother had always encouraged us to apply ourselves and to do well in math and English, but numbers seemed to confuse me, and in the end, I thought I was too stupid to learn. Some of the teachers back then seemed to focus on the brighter kids, leaving us derelicts to struggle in the backwater!

Around this time, I was hanging out with my school buddies, David Vial and Mark Gomez and Scott Johnson, playing basketball, racing slot cars, and flying model airplanes and watching wholesome movies, completely staying away from bad influences, like girls. One time, during the summer after ninth grade, Mark and I rode the bus from Kearns into Salt Lake to go see the movie *Mary Poppins*; it was playing at the Center Theater, which was the same theater we would see *Enter the Dragon* with Bruce Lee a few years later!

Mark and I had just got off the bus and was walking up the sidewalk toward the theater when we noticed a girl about nineteen years old coming our way, she was wearing a "see all the way through" shirt. I couldn't stop looking at her, so Gomez told me to knock it off before we got into trouble, but I remained focused, saying, "I can't, it's my big chance!" Well, this "burn the bra era" babe walks right up to me saying, "What are you lookin' at, punk?" I answered, pointing at her chest, "Them." She responded by socking me right in the eye, and now I was in the catfight from hell! I did my best as Gomez cheered me on, but she finally knocked me down, asking if I wanted some more. I was bawling and holding my eye and said I didn't want any more. The Battle-ax left and went on her way; Gomez and I, we turned around and got back on the bus to go home—no movie.

On the bus, I was still whimpering when one of the passengers asked, "Hey, what's wrong with your friend?" Gomez told them that I just got pounded by a girl, and then there was chuckling. I whimpered some more and said, "Thanks a lot, Gomez."

This was my second girl fight in my life, and I had lost again! The first one had happened in second grade where some little snot named Karen kicked my tail—and for no apparent reason at all. The little meanie (a chipmunk with an attitude) just had to let me have it, and right in front of all my classmates! A friend of mine named Ralph put his arm around me (I was whimpering again) and walked me home, saying, "It'll be okay even if you did get pounded by a girl." I whimpered some more!

Still struggling in school, with my focus on scouting, hobbies, and small part-time jobs, I was angry about my family being on Welfare and free school lunch. Remember, when your peers (classmates) find out you are poor, sometimes you are looked down on; for a kid, it's freakin' embarrassing!

One day while in line for lunch, one of my classmates saw the man taking lunch tickets take my white ticket (most were orange) and slip it in his pocket. This kid made a great big deal out of it. I got lots of stares from everyone around, and I felt very second class! People like me (I guess we were people) were always being reminded of what and who we were! When you would go to pick up your lunch tickets at the school office, it was always a quiet and uncomfortable setting, complete with looks and sneaking around to try to keep others from knowing. I freakin', sucking freakin', hated it! But like a great man (Bruce Lee) once said, "Eating is better than not eating," and I add, "every day of the week!" I should have been grateful for the food, but this business we conducted for free lunch disturbed me and hurt a lot! We also received Sub-for–Santa, which helped to instill an understanding that the world was not completely cold and unforgiving—just like with free school lunch! However, as I saw it, I was taking advantage of others to survive; within me was fire, a raging embarrassed fire burning like a blue supergiant star! That kind of fire can be very self-destructive, especially when burning inside a human being, and at times can be very hard to put out or control.

To make the point in another spectrum, just look at all the movie stars and athletes who were or are so full of fire, sometimes they fizzle out and become lost in drugs, a divorce, or a myriad of other negatives! Me, I was so upset with my predicament after one Christmas, that I completely destroyed a transistor radio that our landlord Dale and his wife Rita had given me. I am so sorry for that action now and am so thankful that there are people in the world who are so selfless, especially when serving thankless ingrates such as me! Because of them and their nature, I am reminded of the song "Amazing Grace" quite often! Dale and his wife were always doing little things to help our struggling family! One night, Dale had come over to repair something (probably a window I had broken) and was told by my mom that I had been bad, so he handcuffed me to a water pipe. Even though everyone thought it was funny, I think it was a small lesson in how life could wind up, that's what I got out of it anyway!

Another reason junior high was a tough time for me is because—we had our bullies to contend with! Back in those days, fighting back against these punks was really the best medicine, but I chose to let those punks make my life hell and chose to take their lumps and find alternate ways to get home, or in other words, avoid them! However, with bullies, there is always the next day, and it becomes a living nightmare! One day toward the end of ninth grade, a kid named Chuck (a terrorist so to speak) walked up and tried to kick me in the groin. I grabbed his foot and lifted his leg straight up (using a karate technique I had just learned) and flipped him backward onto his back, Chuck was so shocked as were all the other onlookers, and he never bothered me again. I was so happy to have this bully off my back because he routinely did something to get me to give up my lunch ticket, I was sick of fear and beatings. I went to a school dance that year and actually had enough confidence to do a slow dance with a girl named Teri Walton. As it turns out, she would be the woman I would marry about nine years later and spend the rest of my life with! Man, I was stepping all over her feet and kept apologizing. She thought I was the biggest dweeb, but girls were scary, and everybody knew (at least in my circle of friends) that to kiss one was to be skinned, cleaned, fried, and eaten!

During our junior high years, David Vial and I were inseparable. We hiked and backpacked together, were in Boy Scouts together, and went to church together and slept out in our yards, we were the best of friends! When we were in seventh grade, I was invited to go to Yellowstone with the Vials on their family vacation. I still remember Hayden Valley and all the buffalo roaming its grassy fields, and fisherman's bridge, which crosses the Snake River flowing out of the park. Most boys back then flew model rockets to one degree or another! One of my and David's friends, Gene Vought, always liked to fly three-stage rockets! It's worth noting that it is very important to align each of the individual stage's fins in a very certain orientation or the launching could go bad! Gene never could get the fins right! One particular time while launching one of these three-stage catastrophes, at the moment of liftoff, all things went to hell in a handbasket, mission control was useless. Instead of flying straight up into the sky, as each stage ignited, it went in its own direction! With kids running around helter-skelter, the final stage went right into the back window of Gene's dad's 1963 Ford Falcon station wagon and set the back seat on fire! Gene got so restricted that we weren't sure if we would ever see him again.

Another time, we launched a rocket from Gene's front yard (parents were gone), which had an unusually large parachute on board! Well, when it came time for the chute to open, it did, but the rocket floated several blocks away and landed in some hooligan's backyard! When we went to get our model, the head hooligan said they didn't have the rocket! We pointed out that we could see someone else with it in the house! So the main hooligan told us that they must scrutinize this device heavily for a while to make sure it wasn't sent by the establishment to scrutinize them, man! And that was that, pecking order established; they kept and/or ripped off our rocket, damned hooligans, damned establishment whoever they are!

Gomez and I were in a band together along with a kid named Doug Mecham, they played trombone and I played the tuba and baritone! I liked the baritone better than the tuba because a tuba's mouthpiece covered half my face and also because kids were always throwing coins, paper balls, pencils, and half-eaten apples into this brass trash can, and it sounded absolutely terrible! On occasion, Mr. Barnett, our "band teacher," would grab my tuba and start rotating it. You could hear the rattling and shuffling of debris headed toward the bell or opening, only to roll out onto the floor with Mr. Barnett looking peeved. Everyone else was quite amused at the junk that thing collected. All three of us had metal shop and wood shop together too! Our teachers Mr. Dalton and Mr. Bond both kept wood paddles that they would spank you with if you got out of line, and you had the option to put your signature on the paddle after the corrective action. However, Mr. Dalton loved to nail ya with a chalkboard eraser too. He would wait till you were totally preoccupied yakking, then throwing the eraser, it became obvious (in living dusty color) who the troublemaker was! We three bandmates were also slot car racers along with my two older brothers—Dennis and Jesse, who was home now and out of the Air Force making a life for himself working for the VA (Veterans Administration) as a physical therapist.

We all raced at Don Browns hobby shop in Kearns or at Bobs Bell Tower Hobby Shop down the road a few miles in Granger or up in Holiday at Hammonds Toy and Hobby with the owner's son, "Hothead" Hammond. These shops were all located in the suburbs of Salt Lake City and not very far apart. Sometimes we even drove to Ogden about fifty miles north to race at Jay and Louis Walker's two hundred and fifty–foot speedway track; the racetracks near home were half that size! Weekly race events were held, and there was various classes you could race in, ranging from stock to super modified, and there was a money pot (twenty bucks or so) to be had if you won your event. Each track had a specific course and required the car owner to match gears, tires, and air foils to track conditions. This hobby required the participant to gain an understanding of model building principles, including electro and mechanical principles. You'd be dealing with soldering brass to spring steel rewinding or de-winding motor armatures, cleaning and truing commutators and armatures, aerodynamics, ball bearings and Oilite bearings, painting, etc. This hobby like most others really taught us kids quite a bit about working with your hands, and you learned sportsmanship too! Surprisingly, these cars in real-life speed are unbelievably fast, the two hundred and fifty–foot track at Ogden had a lap record of about seven seconds; Browns track lap record was just under five seconds, amazing! These days, those times are considered slow; for example, on old man Browns track, with the new Group 27 car, which is cutting-edge, weighing only 2.5 ounces at 1/24 scale, the record is around 2.4 seconds. By the way, the cars of yesterday weighed in at about 8 ounces!

All these shop owners were pretty good mentors and were always giving us things to think about along with the rules of the road that we must follow if we wanted to play. They taught us how to build and maintain our cars, which for me was important, because I bought my cars way back then for ten dollars with money earned at a job that paid me twenty-five cents a day. I have a regret that I must bring up however! While racing up at Hammonds one day, the guy behind the counter, Ken Hansen, let me and my buddy look through a box of parts, he trusted us! In that box was a motor that was the king of all motors. It was called the 24 Single Vulcan and is still, forty years later, the beast! With a little peer pressure naggin' at my side, I stole it! I stole it from a guy who trusted me, a guy who was just like us, a racer! Man, it bugged me from the moment I did it to a point in time just a little while later where I had to go back to our "friend" and turn myself in.

Ken was way too gracious and understanding, he said it was okay and gave me the option of returning it or buying it, so I bought it, and still have that motor to this day; it has helped me win races all over in Utah and Idaho! Years later, Ken wound up getting into the Air National Guard. Full of guilt still, I wondered if he trusted this electrician to come to his office and do work without the fear of being had again! Can you see that every step you take, every track you leave, makes a difference? It does *matter*, it freaking does matter. It always will!

I'm yelling at you, but I'm also yelling at me! I know what is right and what is wrong, so why did I make that mistake, was it peer pressure or fear? They are both stupid reasons, but they are very powerful reasons! However, you eventually have to be responsible for yourself, just yourself. Did you get that? I did, finally!

During this time in my life, I used to go down to the junior high and help a high school guy named Clint fly his big control line model airplanes. One of his planes, called the Flight Streak, had a forty-two-inch wingspan and a Fox Combat .36 cc butt-kickin' engine! Clint was always looking for someone to help him launch this big powerful plane and taught me how to manage my end of the business. He was the pilot and I was the apprentice, "Pit Boss." I had quite a bit of fun thanks to him and was always impressed with how well he handled this crazy-fast combat aircraft! I'll tell ya, when that engine fired up, it would get your heart racing, the high-pitch scream and the smell of burning nitro is, to this day, deeply embedded in my memory! I wound up building the same model a few years later and got to relive some of Clint's and my old times! Many years later, at a time when life was being difficult, I chose to build that model again and got me a Fox Combat .36 X engine off eBay to mount on it! Now my Combat Flight Streak hangs in my workroom on display in honor and remembrance of some of the good old days gone by and a great guy named Clint!

High school was interesting, even though I spent half my school day working at a grocery store known as Mr. J's as a stocker and a bagger; my shift ended at 2:00 a.m. By the way, my life has never been boring, it seems that wherever I wind up, something interesting (good or bad) is waiting around the corner, and a lesson always comes out of it!

While in high school, I had quit taking band and focused on the classes I needed to graduate but did manage to get into an electronics class with my friend Gomez, which seemed to parallel our interest in slot cars! Together we built a TV jammer as our project for the quarter; however, we worked on this device for the entire school year! Mr. Damen, our teacher, would check on us occasionally to see what we were operating on. He never looked surprised, only interested, when we would say that we were still improving our little TV jammer. He would say, "Um, good work," then he'd move on.

One night, I went to visit my brother Dennis and to see his new RCA console color TV! Dennis was very proud of his new TV and elaborated on the great color, contrast, and woodwork! I had brought along the TV jammer for grins and had hid it in my shirt pocket. When Dennis wasn't looking, I turned a knob, then the vertical hold went nuts, and so did Dennis! He pounded on the top of his new TV set; amazingly, this seemed to fix the vertical hold, so I turned the knob a little more, now the screen turned to pure snow. Dennis's face was changing color fast, and he began yelling about a refund and so much for this great TV! I was choking from laughter, immediately Dennis knew that I was part of the problem and demanded the incriminating explanation. So I pulled the TV jammer from my pocket and handed it to Dennis, who examined it and put it in *his* pocket, same as the hooligans did over the rocket.

The job at the grocery store was very important, as it helped pay for my and my sister's school requirements, and it helped me buy presents for family and friends. This job also paid for my first real car, a green-oxidized 1963 Rambler, which had a straight six-cylinder engine and a three on the tree, which is a manual transmission with the shifter on the steering wheel column. I paid fifty bucks for this oil burner that used six quarts a month. It also came complete with a slipping clutch that my bother Dennis said would last a year if I was careful! Plus, the shifter always got stuck between second and third gear, and when it did, I would go out and pry and pound on the linkage (located on the fire wall) till it freed up and put me back in business. The car did get me around, and I used EP reconstituted oil, which costs 15 cents a can to feed the oil-burning problem cheaply. I installed bald truck snow tires on it and told everyone they were drag slicks and cut off the muffler, now I had me a "rod"! It didn't take long before I was walking to work again.

One Friday night, Gomez, a kid named Dave, and I were dragging State Street! At some point during the night, we pulled up alongside a Corvette Stingray at a red light and challenged the punk to a drag race. We both started revving our engines; mine was just billowing blue grimy smoke all over the place, the guy in the 'Vette rolled up his windows trying to escape the stinky cloud. When the light turned green, he laid rubber getting the heck out of there; I dumped my clutch and broke a tooth off a gear! The rest of the night, my "rod" made a loud knocking noise that rose in pitch as the car went faster! I wound up selling this car to my sister Tanya for fifteen bucks, and then she sold it to the junkyard for $25! Interestingly, she and her husband wound up in the used-car business, maybe that car gave them the idea?

While working at Mr. J's, I got to learn the grocery business and how to deal with customers so they will want to come back and shop at your store. The Grocery Store Bible says the first through the tenth commandment is that "the customer is always right," period. One day while bagging groceries for a madam, I placed a weightless package of lunch meat on her bread; she asked me to move it so it wouldn't smash the bread! I replied saying it would be fine as is. She flipped, yelling loud enough that people in Smith's Food King down the street could hear, "I am never shopping in this store again." My throat totally collapsed! Brent, the store manager (a.k.a. Sardonic Man), went after the unhappy customer. She agreed to continue being a customer and might even allow me to bag her groceries again, but it would be the last chance! Brent, being quite an authoritarian (he would have made a great drill instructor), harshly warned me to never make that mistake again because the customer is "the boss," and he was right! Another time around October, a lady customer went to Brent asking if it was possible to buy a pumpkin that didn't already have an eye in its head! It seemed that all the pumpkins had a hole punched into them about where one of the eyes would go. He did find the distressing customer an eyeless pumpkin, but came straight to me afterward because I was one of the supervisors and should know just "who the hell" was putting preoperative eyes in the pumpkins. I actually think I know who the "spear chucker" was and lectured the crew about the problem! But most of our crew were high schoolers and were prone to being irresponsible at times, but not me. However, our manger and his brother were unusually abusive and used fear tactics and physical threats as their main process to get us to perform tasks, so backlashes occurred from time to time.

Eventually, high school graduation came around, and the topic of graduation night needed to be discussed relative to work schedules, so our manager and his brother decided who would get to go to their graduation and who must work that night! A kid named Dave and I were chosen to work (no voting), and if we didn't, we would be fired! I needed the money for my family too much, so Dave and I ran the store that night so long ago. For me, it would be another circle in life left open…

Interestingly, a few years later, our manager and his brother died in a boating accident in southern Utah. Their boat slammed head into a rock wall at high speed, and they were killed instantly!

Change—however it was to be delivered—was coming my way too, like it or not.

4

The Risk

As mentioned, my dad was busy trying to convince me to pack up and leave my mother and come and live with him in Boise, Idaho. I have never figured out why entirely, but something allowed me to pack up my things and leave my home despite my own nightmares and family warnings. Maybe writing this book will help me understand some of this?

Both my brothers told me that if I went to live with him that I would wind up in the military, that I wouldn't be going up there to see new choices for my future like my dad was offering (i.e., college). Interestingly, I had already made one attempt to get into the active duty army! Also, my buddy Gomez and I were both interested in martial arts and had studied kempo karate together at a local Ed Parker dojo during our high school years. Well, we went and saw the Bruce Lee movie *Enter the Dragon* and were blown away and soon after found out that Bruce Lee lived in California and taught gung fu to regular folks like us! So we decided that we could join the army on the so-called buddy system and get stationed at a base near his school in Oakland and our lives would be transformed! Only problem is, the army and their buddy system and duty station promises give way to a basic rule the military always follows, which is, no matter what, military needs come first, period—even over crying mothers and dying cats.

Gomez and I had already been to the old post office in Salt Lake to get checked out physically and take an induction into the army test, which included hemorrhoid exams etc. What a day that was. The recruiting station GIs were just like the ones when Forrest Gump got in the army; they were knockin' us around pretty good army-wise, so we gave it back just as good-wise-guy-wise! Every time some sergeant would call our name, I would answer, "Ya." Gomez would answer with a snide, "Whaaat." We were both told, "You just wait till you two cutie-pies get to boot camp and meet your new momma. This 'ya' and 'what' shit is gonna stop!" Well, in between that day and the day we were gonna be picked up by Sgt. Aguilar, our recruiter, to actually sign up and be sworn in, we found out that Bruce Lee had died several years before *and* that the army had no intention of keeping us together anyway, let alone letting us stay for the rest of life at our chosen duty station in California to become gung fu masters and eventually monastery monks. So on the day that the sergeant came for us, Gomez left town and I changed the address numbers on our house and hid under my bed! Sgt. Aguilar called us a few days later and said that he got the idea—no army for us. However, a few months later, Gomez wound up joining the navy and I packed up my '64 Chevelle Malibu SS and drove to Boise to roll the dice there.

Living with my stepmother, Laretta, and my stepbrothers Lance and Brett was awkward at first; I know it was this way for everyone. I was invading their home and lifestyle, but Laretta tried her best to make things comfortable. She had house rules just like my mom did and she was a fair person. Shortly after I moved in, my stepbrother Michael moved in too. Mike was from my Dad's previous marriage to a woman named Carol, who lived in Boise too! From that relationship, my dad had also created a half sister, Amy, and a stepsister, Sharon—man, I thought, as the world turns!

I have often wondered about Laretta's patience with my father's interjections into her relationship with him. I am sure she had no idea about all the puzzle pieces that he would bring to the table as time went on; Carol probably received similar surprises! I remember Michael, who was a little uncomfortable in asking one time, if his dad was my real father. Both Carol's and Laretta's brood had lived with my father longer than I or any of my immediate siblings had; I wondered what life had been like for them. I already had some idea why Carol was gone and Laretta was in place, we would see again, more puzzle pieces soon enough!

I think Laretta, from the beginning, liked me and did her best to be a mother to me and always set good examples. One night while Laretta and I were talking, she noticed that my grammar was not good, she along the way helped me learn to speak and write better. Laretta had even written a letter to my mother and complained about my educational deficiencies (as she saw it)! My mom gave her some insight into how our lives had been transformed and that we had missed out on a few of the privileges that others may enjoy.

During one of our talks, Laretta asked me if my dad had ever hit my mother. I gave her several examples; she then replied, "Your father has lied to me. He said he has never hit a woman, let alone a wife!" I reaffirmed with, "Yes, and he will hurt you!" I had been suspicious about how things had been for these people as I had noticed a quietness with Lance and Brett—something under the surface, something wrong.

Anyway, it turns out my older brothers were right, my dad started pressuring me to go to school or go sign up for the military, always leaning toward the service. I had no freakin' idea what I wanted to do with my life; I was nineteen and aimless, just existing! So after a trip to Boise State University to see if they taught music recording, which they didn't, my dad introduced me to an Air National Guard recruiter, Sgt. Freddy Pierce. I secretly knew what this might involve. Sgt. Pierce, who was stationed at Gowen Field in Boise, promised that if I passed the entrance exam and joined, he would buy me dinner! Well, I passed the exam and decided to become an electrician. I learned that I would go to the USAF electrical specialist school at Sheppard AFB (Air Force Base) in Wichita, Texas, right after boot camp at Lackland AFB in San Antonio, Texas. So with very few alternatives, I enlisted with the Idaho Air National Guard on October 28, 1975, and would become part of the CE, or civil engineers, unit at Gowen Field after basic training and tech school. True to his word, my recruiter (Fast Freddy) took me to lunch right out to the Air Guard base at the duty mess hall! I was sore and complained to Freddie (my dinner bell pal) because I thought we would wind up at some restaurant in town. Freddy, with a gleam in his eye, said, "We never talked about where we were going to have dinner." *Better get used to it*, I thought!

My stepbrother Mike had already joined the CE unit, which our dad was already part of; Lance and Brett were next in line, but had resisted the pressure through sheer toughness and probably support from their mother. I was scared to go to boot camp clear down in Texas; I had never been that far from home before, nor committed to no turning back either! My dad reassured me that it would be good to get away from all the family for a while to figure me out, and that my drill sergeant would be especially helpful with this issue. I guess I felt a little bit more confident about this decision now, but we'll see.

Also, my dad's volcano had been erupting, just like I remember from long ago, and it was giving me a feeling of dread in my stomach, it made me ill. One day, something tipped him over and got him all wrapped around the axle; he started yelling and screaming and slamming things, and then we heard a loud cracking, tearing, breaking noise. The next thing we knew, he had ripped the family room door in the basement and its casing most of the way out of the wall! This was one of several eruptions that had happened in my short time at my new home, my guts were spinning, and I needed to leave for my own sake!

Next stop, San Antonio, Texas!

I arrived at Lackland Air Force Base in the fall of 1975 with a bunch of other misfits, from now on to be known as Airmen! Technical note: airwomen (WAFS then), by the way, were another matter!

I remember the time being around 11:00 p.m. as our bus arrived from the airport pulling up to a curb in front of some building on base with some airman bossing all the recruits around, just like in Forrest Gump. We were ordered by this bossy yahoo to stand right on some yellow footprints painted on the sidewalk by some knot head, and if you weren't right on the spot, you got nailed. Next, we were all shuffled into a bathroom to do a golden flow test (pee test) to check for substance abuse, and then we were all marched over to our new home, the barracks! Our barracks was a World War II two-story building that had open bays for sleeping. Our outerwear (pants, shirts, and jackets) was stored on wall racks located along the perimeter walls with footlockers at the end of each bunk bed to keep your undies and toiletries in. The whole setup was quite rustic and similar to living in a garage. The structure also contained a latrine on each of the two levels, and the whole place was dimly lit with just a few incandescent lightbulbs on each floor (the military's idea of doing more with less). This is where we would meet our new mom and dad, Sgt.

Mayfield and Sgt. Headd, a.k.a. team chief and team leader, a.k.a. drill instructors or DIs.

Sgt. Headd showed up shortly after we had, introduced himself, and then had us all store anything that we shouldn't have brought from home in a large room that could be locked. One young black kid, Airman Roy McCowen, had bought along a big stereo, a ton of disco records, and a whole wardrobe of disco dancing clothes! Sgt. Headd went ballistic! In a raised, commanding voice, he asked Roy just how long he thought he was gonna be at this training site. "It ain't gonna be forever, boy!" Then the DI stated, "And you ain't moving in with me permanently if I can help it, see! And you ain't gonna be doin' any disco dancing either!" Poor Roy's disco entourage got locked up with a few other nonessentials that others had brought, like basketballs and skateboards and the like. Sgt. Headd then lectured us on the fact that he was sure that our recruiters had made it clear to bring only our body wearing one set of clothes and maybe a jacket, and that was it. Uncle Sam and the United States Air Force would worry about the rest of our needs, and from there on, unless asked to, they would do most of our thinking for us too, period. Well, with all the commotion, I had forgotten the drill instructor's name, and when I started to ask him a question, I had to look at his name tag first. In the gloomy poorly lit room, the last *D* in his name looked like an *O*, so I began my question with, "Sgt. Heado." He glared at me and said, "What did you call me, boy?" I nervously replied, "Uh, Sgt. Heado." He went ballistic again, saying that he was for sure going to physically kill me! Holy cow, I've only been on base a few minutes and I was getting killed already, and there was no way Roy McCowen was gonna be allowed to move in for all time and eternity either. As a pair, we were doomed.

Sgt. Headd then told me, "You better watch your step, boy, you and McCowen!" "Yes, sir," we said in chorus, trying to regain some lost ground. Like it or not, McCowen and I just became targets for the DI's amusement! You don't realize it right then, but this is a pretty normal process at basic training, and the drill instructors in reality have a big job to do in a short amount of time. The next day, we were marched up to clothing issue to receive our various uniforms, complete with footwear and an ink pad/stamp to mark the last four of our social security number in our new duds with. We received work uniforms known as fatigues, which looked great when heavily starched and worn with a spit-shined pair of combat boots! We also received a light tan version that when worn by the black folks, looked extra sharp due to the contrast of colors. We also got socks and underwear and gym clothes and, of course, the official and professional-looking Air Force blues dress uniform, complete with a tie and name tag. This uniform stands out in any crowd, especially when properly decorated! The Air Force has always been sharp dressers, but the Marines are the sharpest in my opinion! Interestingly, there is a reason for the prim and proper military look! Our enemies know that if the military they are going up against shines shoes, presses uniforms, and will keep their haircut that they will follow more strict orders and that they are dealing with a determined well-oiled team! All of this is a reflection of the military's version of the Ten Commandments, esprit de corps!

After we got our uniforms, and a duffel bag to lug them around in, we all went and got paid about forty bucks an airman. This money was to help us buy basic things we needed, such as toiletries; and as such, it was not meant to be squandered. Next stop was the base barber to get a buzz and to pay him five dollars for his work, and then it was on to the BX (Base Exchange store) to get those basic necessities like bar soap, toothpaste, razors, and such. Interestingly, as we loosely marched along, we heard the catcall "Rainbow," and it was directed at us! We were rainbows due to our multicolored civilian clothes and would be misfits until we got into our uniforms, which were now on our backs in a duffel bag.

As I said earlier, the drill instructors have a lot to do; they have to get a bunch of joes from all over the country to walk, talk, and think alike and become reasonable young men who can follow orders. Unfortunately, some of these joes just never fit in! There was a kid from Illinois named Wilder, and from the very start, he went against the authoritarian nature of our DIs. One day while out on the drill pad learning our left-face and about-faces, Sgt. Mayfield, our team chief, seemed to be homing in on Wilder about not paying attention. Wilder, like I said, had been abrasive from day one and challenged the sergeant! Sgt. Mayfield brought Airman Wilder to attention and, in the DI's indomitable fashion, got nose to nose with this kid and tried his best to get him to take a swing. Wilder thought about it, but in the end became the subordinate that he was. Eventually and unfortunately, Airman Wilder washed out and became a member of the detention flight, which ain't good! DTs (detention flight troops) run in formation everywhere they go, and when at chow, they have to ask permission to pick up their fork, ask permission to put food on it, ask to put the food in their mouth and eat—get the idea? At the end of the day, they are taken to their quarters, which is a jail! They must change their errant ways by force or they wash out completely, and who knows what their fate becomes…

Sgt. Mayfield was with us about half the time, and his focus seemed to be set on certain drills and inspections, he was supposedly a retired Marine and definitely fit the profile. Sgt. Mayfield had special requirements that he would impose on our barracks guard, or House Mouse, which is what Sgt. Mayfield preferred to call him. The House Mouse was one of us airmen assigned to watch the barracks for a twelve-hour shift. We guarded our own barracks 24/7. Well, when Sgt. Mayfield was on duty and you were the Mouse, you better have his coffee ready at 5:00 a.m. sharp, and by damned, it better be a certain temperature, or he might put you in a footlocker and shove you out a convenient upstairs window.

One morning, Sgt. Mayfield came into the barracks about 4:30 a.m., nice and quiet so as not to wake up his sleeping girly-mans and went straight to his office to get his cup of coffee. All of a sudden, there was a crashing and a combined yelling that just scared the hell out of everyone! It sounded like there was a mad bull in the kitchen, smashing pots and pans, flipping over tables, etc. Then Mayfield started screaming bloody murder, "Where the hell is the House Mouse," ending with, "I burned my lips!"

That's just great, Mayfield burns his lips and we get a surprise footlocker inspection as a result, then we had to scrub and wax our barracks floor, clean our latrine beyond 136 percent perfect or else! Then we got relectured about coffee temperature, and then we did an FOD (foreign object debris) walk around our barracks, weeding and picking up cigarette butts. The whole time we were being hounded by both our DIs, Dr. Jekyll and Mr. Hyde. Happy days just fall all over new recruits. We learned that Sgt. Mayfield was also very concerned with timeliness; unfortunately, Airman McCowen always seemed to be last! One day we were marched back to our barracks to get dressed for PT (physical training). (We were allowed five minutes.) Sgt. Mayfield was already threatening McCowen! Boy, he said, "If you are the last one back out here to form up, I'm gonna stuff you in your footlocker and push you out one of them second-floor windows. Do you hear me, boy? Go!" he commanded! I think McCowen truly believed the DI would do what he said. Minutes later, McCowen, still about third from last, fell in, so he was safe. The drill sergeant complimented Airman McCowen on not being last but then asked Roy if he realized that he had forgotten something. In his haste, Airman McCowen had put everything on except his workout shorts; there he stood at attention with his jock strap, running shoes, T-shirt, and a bare bottom! Everyone started coughing, which was a way to hide a laugh! The drill instructor told us to knock it off, that he knew we weren't all sick! Roy was allowed to go put on his running shorts, minus the footlocker routine.

The DIs knew when to give and take, and we would learn to respect them for this wisdom. One day a few weeks later, our flight was at the chow hall waiting our turn to go in for lunch while our DI had gone inside to do something. While we were standing in formation at attention, a flight of WAFS (Women's Auxiliary Ferrying Squadron) marched up and was brought to attention by their DI; they were new recruits, thus, "Rainbows!" One or two of our guys started catcalling "Rainbows," and we all started laughing. Immediately if not quicker, the female DI jumped on her broom and zoomed right over to our flight and ripped us up like a human buzz saw including a slew of insults and then ordered us down on the ground to do pushups till she decided we could stop. Meanwhile, our DI sergeant, Mayfield, returned and seemed to approve of us girly-mans getting whopped by the wicked witch of the west! Holy cow, I had no idea a girl could be that mean, and all our "good buddy" Sgt. Mayfield had to say was, "That'll teach ya!"

We were learning to become disciplined, responsible, intelligent, acting human beings and were now part of a formidable and highly respected organization and would spend lifetimes earning the right to remain!

Back on day 2 during our initial indoctrination, the recruits all received 341 discrepancy forms, which is a way for the DI to discipline unruly airmen! How they are used is, if the DI catches someone committing some major offense (i.e., looking at an airplane flying by while marching etc.), he will take one of the three you are carrying in your right front fatigue shirt pocket! Typically, after he has taken all three, he may have you report to the kitchen at the chow hall and complete a shift or two of washing pots and pans, which is done by hand and can last about sixteen hours, but you eat good! So it pays to be smart and try to act like you don't like it.

In boot camp as in life, teamwork is the key; you must wrap your head around the idea of unity, or the sign of one, or united we stand divided we fall. It's so important to learn this principle that even when we went to the chow hall, no one sat to eat at a table until every space available had a body and his or her food before them. Marching is a great way to see how a flight is working as a team (i.e., when a group of soldiers is marching in unison and driving those combat boot heels firmly into the ground with all body motions in unison). It's intimidating, as it is a single sound of thunder and authority and makes the DI and the troops proud!

While in basic training, we learned about one another and our individuality and how that interacted with our becoming a team. Like all other flights (Air Force term for *unit*), it seems that we were from all over the country, we all had our own beliefs, values, religions, and ways of living life! But now, we all belonged to the US Air Force, who had the only way to see things, all things in the universe even; nothing was left to chance, like I said, unity! We were taught how to walk, talk, and be fit in mind, body, and soul; we were taught manners and discipline and how and why to be clean and organized. All this was given to us with plenty of TLC (tender loving care) from our new parents—not surrogate either. As they saw it, they were the ones who gave us life! None of this means that we airmen didn't have a little in house fun now and then. Most of us were teenagers and prone to practical jokes, smart comments, etc. Us guys who bunked upstairs were all-night owls and loved to talk relentlessly, which kept the guys downstairs awake. Those downstairs bums would cry and moan about the 5:00 a.m. wake-up call, which came every morning and was only a few hours away. Sometimes we would be real quiet till about eleven, and then one of us, who shall remain nameless, would quietly go to a place on our floor that was right over a downstairs bunk, belonging to a kid named Jackson, and pound the floor with combat boots, then slither back to bed. Jackson, a six and a half feet tall nearly boneless farmer from Kansas, would leap up the stairs, bypassing half of them, to catch the filthy animal that did this noisemaking. Jackson would reach the top of the stairs looking like an Irish setter with distemper, only to be stunned because all was quiet. Then he would blame Airman Cervantes (one of my bunk mates), saying, "I know it's you, Cervantes, and one day, I'll catch ya, then I'll tell the DI and he'll pull one of your three-forty-ones." Everyone bit their cheeks, it was hilarious, we did it often and just for fun!

The Chicago stud muffins were always bragging about how much money they were making before they were inducted into the Air Force to become poor. They also yapped about all the women they had fun with too, that is, until some young virgin they had made fun of reminded them about the ugly and gruesome effects of VD (venereal disease); after that, the room was dead quiet! None of us kids really knew much about those realities of life, we only had old wives' tales to go on, and those tales were scary enough and wrought with questions.

Those guys downstairs were right though; the nights passed in a blink, and we really needed our sleep. You see, between 10:00 p.m. and 5:00 a.m. was our time; mom and dad weren't home, and we felt a little bit free. But at 5:00 a.m. sharp, you would hear the barracks security guard (the Mouse) let the DI into his house, then you would hear the taps on his boots clicking on the hard tile floor. Now the day was his to whittle and hone us up some more and mold us into respectable military men who would be able to continue to learn and grow and be a useful part of the United Sates fighting forces! Again and again we were taught teamwork and respect and how to win, and we were taught to face our problems and how to recognize problems in others and how to help! Over and over we were constantly reminded about esprit de corp and just how very important that term is to our whole being and our country, the United States of America! We didn't realize it then, but we were being prepared for life in a way that we would not have gotten, at our homes far away; this was a great thing that my mean dad did for me! All this was free of charge, except the fine print says you gotta give up a little bit of your time and freedom. Also, if you fail boot camp, you may have to go through the entire course again or be removed from the Air Force altogether, which might be a very bad thing for your future!

Some of us would stay with the service for the rest of our lives, and others would move on to other ventures. However, I believe to a man that boot camp left a lasting impression on each one of us and helped us become better and more productive people. When and if you leave boot camp, you are a honed individual and probably can swear better than ever. You are also more confident than you have ever been and are ready for your career field training at an Air Force technical school somewhere in the United States. Me, I was headed for Sheppard Air Force Base in Wichita, Texas, just before Christmas.

Not everyone in your flight at basic training goes to the same technical school! Only one other airman, Curtis Hoyt, from my flight would go to Sheppard AFB with me and specifically to the electrical specialist school. We lucked out and got to bunk together in a "two men to a room" dorm that housed several hundred airmen with community showers and restrooms, which we maintained. Typically these technical training sites are a little more relaxed than boot camp; there are no drill instructors hounding you about how to fold your Skivvies, you are past that stage. Yes, there are still military courtesies to follow, and you have to maintain your military bearing at all times, but it's natural to you now and you feel good about doing it!

Military people are set apart because of their military bearing and should always exhibit a high level of professionalism! Yes, we can, at times, let our hair down as it were, but there is a reasonable limit to everything in life, especially with the service and what it represents!

I remember as our bus was pulling onto the base, Curtis yelped, "Hey, there's a McDonalds right across the street from the main gate"; we were both licking our chops! Air Force food is good, mind you, but Macky D sounded like a great change of pace. After we were "in-briefed" about all the important stuff like where our billets (living quarters), schools, and dining facilities were, and duty hours, the time after that was ours. Once our in-briefers cut us free, Curtis and I headed straight for McDonalds, where because I said I could, Curtis bought me five Big Mac combo meals. I ate them all including the fries, and then just about croaked! Curtis laughed like crazy at his dumb and sick friend! Despite a pretty busy school schedule, we had time to build and fly control line–style model airplanes and had loads of fun doing that hobby! We had a buddy in our electrical class named Chris; he had never flown model airplanes but wanted to. So we all went to the base hobby shop where Chris found a plane he wanted to buy. It was an advanced model known as a Stuka, he referred to it as a Stuki! We tried to talk him into a trainer, something he could handle, but it was the Stuki or die! So when we got it all set up back at our dorm grounds, we suggested that he fly it over the lawn; Chris balked saying he would take it off and fly it over the parking lot, we grumbled.

So I started the engine, which started shrieking and screaming and scared old Chris to death! Now he did not want to fly his Stuki at all, but Curt told him, "It's yours, you fly it." So with a nod from Chris, I let the airplane go! From that point on, no place was safe, the plane would bank toward the ground and a group of onlookers, and then he would yank on the strings forcing his Stuki in another direction. It was total hysteria, nobody knew which way to run! Finally the red gods of fate sent the plane into a curb, and that was the end of the Stuki! Pieces were everywhere, Chris half in tears could only say, "My Stuki." We were sad for our friend and did our best to repair this dead plastic horse, but it never flew again.

The training at Sheppard AFB we were receiving was really top-notch (I would realize this later in my career), and many of our instructors were civilians! They did a pretty good job preparing us for our home stations where we would receive loads of OJT (on-the-job training), including additional upgrade training, mostly in the form of CDCs (career development courses).

The topics were fairly broad in scope and quite similar to civilian electrical apprenticeships, with a few differences due to duty requirements! This part of my training would last through April 1976 (about four months), then I would return to my home station in Boise.

During Christmas vacation, Airman Adrian Andrews (from Hawaii, and a classmate) and I decided to try some "Texas" beer out! We got pretty sauced up pretty fast and stumbled our way over to the airmen's club to join the rest of the folks who hadn't gone home for the holidays. Well, because we thought we should, we started trying out some Black Russians, which, when mixed with the beer, boiled our brains, but we were tough, we were in the military. After we had raised our courage level a might, we climbed up on our table and challenged the whole club to a big brawl; the crowd told us to shut our pieholes and to get off the table, so we told them they were afraid—it went back and forth! Finally we gave up on these scaredy-cats and climbed off the table—mostly because we really had to pee; when we made it to the bathroom, it looked different than before! Presently a girl came out of a stall and told us to get the hell out of the ladies room! I told her that we can pee in here just as good as she could! Next thing I know, she is coming at me with her fists doubled up and then I could hear thuds on my head; a couple of thuds later, I was on the ground with her standing above me waiting to give me some more "What for!" My partner Adrian started telling me that I had to get up because the Air Police (a.k.a. Sky Cops) were coming; I said that I couldn't because the witch had punched me out. Then I woke up to a captain talking to me in some office; he told me that I couldn't come into this airman's club and cause trouble even if the girls do beat me up! So the command quarters security guards (our dorm guards) came to take me back to my room and put me to bed. I was your typical drunk; I refused to walk, so these guys threatened to pound me, I told them, "Go ahead, a girl has already beat me up!" They thought that was pretty funny and dragged me home and told my bunky, Curtis, to keep me in for the night. The next day was terrible, my head hurt all day! I really have never been a drinker, and this turned out to be a pretty good lesson, especially when it comes to getting into girl fights. I wasn't a fighter nor much of a lover! I was just some kid from Idaho and had just lost his third and final girl fight!

Eventually, I made it back to Boise safe and sound from Texas and that very different experience and was better off for it; it was the spring of 1976! I also got settled into my new guard unit Civil Engineers; my supervisor was Leonard Brown, a wise old real world-master electrician and senior master sergeant. Along the way, Leonard taught me lots about our specialty; and years later, I realize just how capable Leonard was, especially in passing his great skills and knowledge on to others. However, I would be moving on; a long-term place in that unit was not meant to be!

After a few weeks of adjusting to civilian life, I got a job as an electrician roping mobile homes (a term used when wiring residential properties) and worked for a man named Larry. For transportation, I had bought a 1973 Honda CB500 road bike from my new Air Guard supervisor's brother, Wally, who was also in our CE unit! Wally and Leonard were planning on buying the brand-new Honda Gold Wing GL1000, so Wally needed to sell his beautiful 500. The 500 was in literally perfect condition, not a scratch or smear of grease was visible! Wally told me that he would not ride the bike if there was a cloud in the sky and that it got a wash and a wax weekly! Wally then told me that he would sell this fine machine (his baby) to me only if I promised to give it the same care, I promised I would. Wally then forlornly made a remark, "I hope I don't regret selling this bike." It wasn't a month later when Wally and Leonard were out riding their beautiful new aqua Gold Wings, and as they were rounding a turn, a cantaloupe-sized rock was lying on the road! Leonard missed it, but Wally hit it with his front tire, forcing the wheel to the right, opposite the direction of the turn they were making! This caused the handlebars and wheel to be forced into a full right turn with enough energy to cause the handlebars to break through the fiberglass fairing that supports the windshield. Now, the bike was locked into full right turn! As you can imagine, at 60 miles an hour, the bike just started flipping sideways. Immediately, Wally was launched into the air! Leonard, watching in horror off his side-view mirror, saw some of the catastrophe unfold; he saw his bother launch and then hit the road! The bike was destroyed, but Wally managed to live; however, his injuries changed his life for good! The cause was that the fairings position was in the way of the full travel of the handlebars, which needed to be set back several inches for the proper clearance to occur. So an unseen danger was present until somebody got hurt, and then it got figured out and corrected. I went to see old Wally one day, and he reminded me about what he had said about regretting selling the 500. Interestingly, my dad cosigned with me on my bike, then turned around and paid for my three stepbrothers' bikes, which made them embarrassed. They came to me to apologize, but I told them, "Look, I am the oldest and am out of school etc. and should be more responsible for these sorts of things!" I hope that helped them deal with the uncomfortable way they felt, and it did make me feel more responsible for myself—plus, you gotta get there sometime!

During my first week working at the trailer factory, while on a work break, I met a few members of a chapter of a known motorcycle club. Many of the charter members worked here at the trailer factory! Their leader, Farly Wadsworth, said he noticed that I rode a motorcycle and told me that I could park my bike near their choppers. However, I wasn't to get the idea that I was any part of their gang! Farly then said, "The same goes for lunch and breaks." I could sit near them in the break room, but I was in no way to think that I was part of them! To me, this meant no blood initiation and no club jacket, dang nab it! They looked tough and ruthless, but really were nice enough and worked very hard doing a nasty job, which was insulating with unbacked fiberglass, the entire mobile home! I thought it was funny that they were outcasts, and I considered myself to be sort of an outcast; but here, I was an outcast to the outcasts, really the bottom line! I was also about to find out that I needed far more skills than I possessed to have the position that I was hired for, which was to replace their previous journeyman electrician. At this time in my life, the only thing qualified about me was my hopes and dreams.

Meanwhile, the towering inferno volcano (my dad) was erupting regularly, making my return to Boise tense and unsure. Through the summer of 1976, my dad got wrapped around the axle again about various topics, granted a few of them might upset any other ordinary person, but several times he went ballistic!

One day we were all building a cedarwood fence around the yard. Laretta, who was helping, noticed that as we went, it was becoming increasingly unlevel. My dad, who has a degree in mechanical engineering, did not like hearing that the project he thought he was in charge of (but Laretta really was) had an issue with level! A serious debate began, which became very heated on one side! Loretta maintained her cool, but my dad would not accept that his fence was unlevel; suddenly he raised the hammer that was in his right hand with the look and intent of hitting Laretta! My three stepbrothers and I froze, we were scared and in shock! Well, 110-pound feisty little Laretta grabbed a ten-inch crescent wrench from a nearby bench, raised it, then faced Goliath, saying, "Come on, ya big SOB!" The two combatants stared each other down, all was silent for an eternity, or so it seemed. I knew that I may have to step in and save my new stepmom; I couldn't let my dad hit her with his fists or feet, let alone a hammer. Fortunately, somehow the situation diffused, and the talk became a little more reasonable. Silently I wondered just how much Laretta and her boys had endured up to now, let alone Carol and Mike and his sisters (my dad's previous responsibility). I have never heard why my dad and Carol separated; however, I am sure his infidelity and his strain of torture tactics lie at the root! Amazingly, Laretta, to this day is still married to my dad! Recently, during one of our mother-to-son talks, she lamented, "You get so far along in life [they are both in their eighties] that you just try to stick it out and accept it," stating that she had been through worse! That comment from this strong, compassionate, and loving woman sounded painful to me. I guess, for her, it's too late to run… I wonder if my dad, in the twilight of his life, realizes anything about all these episodes and all the people he impacted—for example, the ones he should have cared for and loved the most. However, the next eruption would finish me off.

One fine day, one of my stepbrothers talked back to Laretta! You *never* did that with my dad; it was the equivalent of causing a grizzly bear to become fully enraged! Suddenly, I heard running, yelling, and screaming and then a large crash! In an effort to get to the mouthy kid, my dad "cleared the road," destroying a five-foot-diameter glass table, then there was more yelling, death was in the air! My guts tightened into a twisted knot and churned till I thought I would throw up! I yelled in frustration, anger, and fear, "I'm leaving!" I immediately called my brother Dennis and asked if he could come and get me and take me back to Salt Lake City. Dennis told me to get my things ready and that he and Jess would be up in a few days to move me back home! The scariest thing I had to do after that phone call was to tell my dad about my plans, this was like facing a judge for the murder of another human being! I expected a brutal bashing or a physical beating, but all he did was his best to convince me that I was making a mistake and to stay! The look on Laretta's and the three boys' faces were "why do you have to leave?" They knew that I was breaking out of the prison they were in.

In the meantime, my dad informed me that if I left, he would not let me take my motorcycle because he cosigned for it etc., big damn deal, I thought. So I put it up for sale at the local Honda dealership to shut him up. However, I still had to make the payments, and I would do this because it was the responsible thing to do, plus I did not want my dad having an "I told you so" to use on me. A few months later, a new boss would cosign on a loan for me, and I would retrieve the Honda 500, relieving my dad's worries about getting stuck with the bike. Several days later, my two older brothers showed up to help, and we headed home, back to where I belong.

Just a thought: At this point in life, deep down, I really didn't see much value in my life, I felt lost, I seemed to be taking more than I was giving. To me, I didn't matter, why else would things go like they had! Inside me, however, there were questions along with fire and some hope that the military had instilled. Mean abusive people like my dad may never know or care about their impact on others' lives, they just wade through people like they are so much garbage. Amazingly, there are people who are unbelievably worse! I'll tell you what though, I often reflect on the things that I have said or done, I too have regrets.

I am always amazed when I hear about abused people surviving their epic trials and then succeeding in some way in the aftermath with the odds against them. They must have very badly wanted things to be better and believed that it could be despite the pain and suffering. Simply put, they persevered and had faith that things would improve, no matter what! My childhood story seems so commonplace that it makes me wonder why I would tell it. Parental abuse, divorce, living on welfare, or being out on the street is an everyday fact of life! Some of us just find a way to survive one way or another, and some don't and simply fade away. It had occurred to me that my life seems to be hell-bent on, and designed to be full of, brutal trials. It's like I'm being tested for some reason physically and mentally with great harshness. Is the Creator trying to see at what point I will fold and turn in my cards? I kept thinking that I need to keep going and see what is around the next bend. Interestingly, this very process of curiosity has helped me get on top of mountains in faraway places, finish projects, and I think, ultimately, has kept me alive!

5

The Repeat with a Plus

Back in Salt Lake City, I moved in with my mom; we were glad to be together again. I insisted that I pay my share of the household expenses as she was now living all alone and it was a task for her to survive. My helping took a bit of pressure off her, and I know that she appreciated it. My mom (Marge) was now working at Deseret Pharmaceutical, in a clean room making critical parts for various surgical needs. I am sure that it was good for her to be out of the house and having a life with friends and coworkers! I had recently completed the transfer process from my previous Air Guard unit at Gowen Field in Boise to my new unit in Salt Lake City, the 151st civil engineering squadron, and would be in the electrical shop, which was run by SMSgt. Harry W. Allred, a very capable master electrician, just like Leonard back in Idaho.

The first person I met in my new unit was Chief Mont Gunderson, who was really a senior master sergeant, but his nickname to some was Chief! So there I was walking down the road toward our building, which was an old bomb storage facility converted to office space. As I got close, I could see this big face pressed against the main entrance door's eighteen-by-eighteen-inch window! I remember thinking, *Wow, that guy's face is big, it nearly fills the framework.* As I entered the door, Chief Gunderson, who was as big as Jason in Halloween, was eyeing me suspiciously, instantly barked, "You, soldier." I replied with a long "whaaat?" The Chief then asked what I was doing with chukka boots on, which were ankle-high combat boots issued to me at boot camp. Lacking for a better at-the-spur-of-the-moment comeback, I stated, "Well, I guess I'm just wearin' 'em, aren't I!" Chief Gunderson barked, "Don't you smart off to me or you'll be out of this unit." Heck, I wasn't really in the unit yet, so I said, "Don't threaten me with sleeping in every weekend of the month 'cause I'll take it." Mont replied, "Fine, I'm going to your supervisor, mister." I had no idea who that was, but I got to meet SMSgt. Harry W. Allred pretty quick. Harry told me to stay away from Mont and I'd be better off, that he hates everybody especially new recruits, and is interested in kicking everyone out of the unit! Later that year, I had another run-in with Chief Gunderson; he wanted to kick me and a buddy out because we didn't like the issued socks because they crawled. You couldn't take two steps without them crawling down inside your footwear. But Mont said, "You'll wear them and like it, or you'll both be out of the unit!" Another time, I folded up and then stuffed my sleeping bag into my flight bag (which is a rectangular duffel bag with a zipper) instead of rolling it up and jamming it into the already overfull flight bag. Mont saw what I was doing and went ballistic, stating that the sleeping bag is not rolled up, so it's not regulation! "Roll it up or you are gone," he demanded. I pleaded my case. "Roll it up or you are gone" was the final verdict from the judge, jury, and executioner Chief Gunderson. Personally, I would have hated having Mont for a prisoner of war, I'd let him escape unmolested!

"Montpelier," as some of us liked to call him, really played no favorites and hacked on everyone, including officers, so I wasn't alone when up against the wall. However, in so doing, Mont had to understand that to live by the sword, you must also fall victim by the sword! This means that Mont would be the recipient of a few practical jokes; and there is no better time to seek revenge than when on a deployment—as you will later see.

At the time, I had no idea just how influential the Utah Air Guard would be in forming me into a responsible human being and soldier! Nor did I realize then how much effort SMSgt. Allred (Harry) would put into turning me into a knowledgeable and skilled electrician.

Being in a CE unit fit me well however! I have always enjoyed building things and taking them apart too, just ask my brothers. They would give me a slot car or model airplane, on the premise and promise that I would not take it apart; I would agree, and a short time later, my curiosity forced me to disassemble the item! I was a curious mechanic, and this curiosity would later lead me into very technical terrain regarding electricity and the venture of mountaineering and its related equipment.

During the 1970s, the Air National Guard hardly seemed to be on the radar. Back then, we were supposedly a backup for the active duty forces, that is, if things went really bad in a war! I suppose, if things went so bad that you were calling up the citizen soldier to come and fight, the possibility of adopting a new cuisine and language was imminent!

Otherwise, homeland security (e.g., floods, fires, snow removal, and crowd control) was a responsibility of ours, that is, if the president and governors saw fit. Our unit deployed in and out of the country quite a bit though, building, repairing, and maintaining things for the Air Force and other branches of the service. And of course we were also busy doing the same things at our home station in Salt Lake City. Our unit excelled at these various tasks, largely due to the fact that a major percentage of our members were skilled tradesman in the civilian world. As a result, we were given unique assignments all over the globe. Sometimes these assignments were humanitarian in nature, and we were able to help needy people have a school or a hospital, maybe even a church. When working for various branches of the service, be it our own or an allied nation, we seemed to focus on building and/or remodeling barracks, offices, and shops, and medical facilities. We even laid asphalt for roads and parking lots and poured concrete pads too. We also built server rooms, shooting ranges, runways, and their lighting systems as well! Many of these projects had to be engineered when we got there by our unit technicians and engineering staff, which means there better be some pretty capable folks along when this need arises! Many of these tasks could become our responsibility in the event our country went to war or if we were back filling at some base while its members were deployed. Again, it's a lot of responsibility! Sometimes you might deploy to a site that will require the Bare Base Process to be used; in other words, you are setting up tent cities and, within that infrastructure, creating substations and water purification sites, medical, laundry and dining services, maintenance shops, runways, etc. When it came to the runways, the electrical shop, besides the lightning, set up the glide slope indicators and the MAAS (mobile aircraft arresting system) that tail hooks specific aircraft to get them stopped in time after landing! In the event of an attack on the runway, there was a triple R (RRR, or Rapid Runway Repair) team that responded and was comprised mostly of heavy equipment operators, electricians, and power-pro people! Our job was to reestablish the runway after it got bombed. The team had to be able to accomplish the task in two and a half hours max, which at one time was the turnaround time for bombing runs! If you were in a situation that required RRR teams to deploy, your heavy equipment and EOD (explosive ordnance disposal) folks scraped the runway for ordnance, then the site surveyors would establish the new runway centerline. Next up, the heavy equipment operators, a.k.a. Dirt Diggers, started refilling and compacting the huge craters, preparing them for the huge aluminum AM-2 mat that was being built nearby, which would soon cover the dirt/road base–filled crater! AM-2 matting is aluminum and is comprised of 2' × 8' interlocking planks that weigh 160 pounds each, a few 2' × 4' planks, plus the leading and trailing edge pieces that must be anchored to the medium! Building a mat to cover an 800-pound bomb crater is seriously hard work when done in fatigue dress, let alone chemical/MOP (massive ordnance penetrator) gear! The AM-2 mat could be built wrong too, with its leading edges (which are low profile to help airplane wheels safely enter the medium) winding up on the wrong side of the mat; also, the mat could be built upside down. Many of these these occur while you are under stress; most of these things are critical to everyone's safety and, if done wrong, must be reaccomplished! Meanwhile, the electrical and power-pro shop would be focused on the lighting and arresting systems. The lighting system is very dangerous as it operates at 2,400 volts and has taken a life or two because someone wasn't paying attention while performing maintenance.

My first exposure to all of this was in 1977 at Tyndall Air Force Base in Florida while on an exercise called BRATT, which stands for "base recovery after attack," pretty nifty eh! When we arrived at Tyndall, the mock war was already on. We had just flown from Salt Lake City on a freezing-cold C-130 cargo plane for seven hours to 100-degree Florida. Now we were being shot at and had thirty-six hours to set up and maintain a base complete with potable water and a functional runway. Well, there were things that did happen in that thirty-six hours and there were things that didn't! Our water came out of the ROPU (reverse osmosis purification unit) water treatment system black, and our runway didn't get completed. But we did get the tents all up and lit, and we did eat "sea" rations and did get bit by every bug that wanted to bite something! Turns out that this routine scenario needed some tweaks! So the Air Force eventually turned BRATT into a five-day program called Silver Flag, which included three days of classroom training with one day of practice war, plus one day of war that counted for a grade! If your unit failed the day when it all counts, you had to come back that year to be reevaluated. If you failed again, your commander and/or unit would be in serious jeopardy!

During the eighties, we deployed to Alaska in late summer to build armories at remote sites for a division of the army known as the Alaskan Scouts! These remote sites are scattered all over the mainland and the Aleutian Islands and had interesting names such as Kwigillingok, Mountain Village, and Kivalina, to name a few. The Scouts members were primarily Native Inuits and Eskimos who were signed up to be, near as I could figure, border patrol. These armories we built were prefab single-story structures, and we put them together with simple manpower and hand tools! We had no heavy equipment to lift the trusses or the 800-pound glulam beams either. The days were long, the food pretty good, and we got to sleep in "ages old" fish-cleaning shacks in sleeping bags on cots, and it was high adventure. We were so lucky to be living and working in these remote areas of Alaska and even once in a while got to fish for salmon and arctic char in the rivers near the villages! Life and the company were pretty good up there in the far North during those days.

It was while we were deployed to Kivilina constructing an armory for the Alaskan scouts that Chief Gunderson got some warranted "payback." A friend of mine—Bill, from Kearns, a new recruit to our unit—had come along on his first deployment to help me wire the armory. I had warned Bill back in Utah about Mont's nature. Stay out of Mont's way and everything will be fine, I told him; giving a few examples, Bill was sure I was exaggerating! Well, one day, Mont was in the main room of the armory screwing some weightless metal channel to the wood trusses. He needed to create sixteen-inch centers for attaching Sheetrock and was using an AC (plug-and-cord operated) electric screw gun. As the chief worked, his extension cord became tangled around some 2" × 4" boards lying on the floor. Well, happy-go-lucky Bill happened by, so Mont asked, "Hey, Bill, would you untangle my cord." Eager to please, Bill said, "Sure, sure, Sgt. Gunderson, you bet!" However, during the process, Bill disconnected the drill from the extension cord (kinda had to). Oh my hell, Mont threw a freaking fit bellowing, "Look what you've done, just look at what you have done, my drill is unplugged, and I am standing here holding all this metal, and now I can't work!" Bill was stunned by the region's newest storm, better known as category 4 Hurricane Mont! So Bill stated that he could see that this was a real problem, a dilemma, and that 10 ounces of tin is so heavy? Well, Hurricane Mont became a category 8, and Bill, well...Bill brought me lots of cups of cocoa repaying me for not following my advice. Later on that day, Bill told me, "I sure would like to pull a prank on that yahoo. Wanna help?" I replied, "Sure, what do I do?" Chief Gunderson (Montpelier, don't forget!) had an alarm clock. So we snuck into his room every day from that point on and set his time ahead two hours and set his alarm forward two hours. Mont's eye whites became pretty red in the Land of the Midnight Sun. Sadly, we never heard a word about the clock caper, not a word. Darn that Mont!

During the mid-eighties, our government sent my guard unit on a humanitarian mission to La Paz, Bolivia, a beautiful country located in central South America. Our job was to renovate and modernize a one-hundred-plus-year-old hospital that was quite large, but so is La Paz, which had at the time six times the population of Salt Lake crammed into a similar-sized space. We arrived dressed as civilians and needed to blend in for our own safety, plus it was election year. We had been told that Bolivia had more governments voted into office than their years of existence? We would stay in a three-star hotel named Gloria and would alter our route to work routinely and stay out of trouble. Since there was a machine gun complete with an operator on every corner in this martial law–ruled nation, I think we were more than happy to mind our p's and q's. We were also told by the embassy that at this time, there was only one ambulance for the entire city and only three freezers for corpses in this short-staffed hospital. The bodies in the freezer room were stacked up like so much cordwood. Interestingly, as we drove down to La Paz from the Alti Plano (High Plains), where the airport sits at 13,500 feet (the highest in the world), we were paralleling a glacially fed river that runs through La Paz, it was changing from milky white (due to glacial silt) to a point within the city where it was black! Our hotel water came from this river but was purified; however, I still remember being hit with small chunks of something when I showered. Nevertheless, most everyone eventually contracted and suffered from amoebic dysentery. The rule here in La Paz was "Don't get hurt!"

The scope of the remodel was an immense undertaking and focused on upgrading the entire electrical and plumbing systems to current US of A standards. The project also included replacing the very dated x-ray equipment, which reminded me of the old pedal-type Singer sewing machines with wrought iron frames; most of the other medical equipment and tools were being replaced too. We would also upgrade their commercial kitchen and their commercial laundry, thus attempting to bring them out of the Dark Ages. The current electrical system looked like it came right out of a Frankenstein movie. The conductors were draped in the open across the walls and would connect to open knife switches, rendering all energized parts exposed (this was the Frankenstein part)! The laundry was even more austere. All the bedding and clothing for the hospital were washed by hand in a "fourteen-by-fourteen-foot by several-foot-deep" stone tub! The water was heated by a very contrived man-made heating element. Seriously, there were two wires leading into the tub from some source that connected to a chunk of metal with some Bolivian guy (with a serious look) wearing rubber boots and gloves, tending this oversized water heater. Our plumber, Harold, found out the hard way why the tub tender was wearing the rubber goods, as he got shocked with 200-plus volts while standing in water near the tub and was seriously injured. We would also have to hand dig deep and lengthy trenches for the electrical and plumbing installations. Our government paid our so-called coolies one US dollar a day to perform this very difficult task! "The Bolivian people are very hard workers and were always happy to help." Incidentally, a person could eat for a US dollar a day back then, it wouldn't be steak but you could survive. This is how things are in third world nations, and it can even be far worse making this place seem luxurious! When we left Bolivia, we gave our civilian clothes to these humble people, who broke down into tears of gratitude, giving us big heartfelt hugs. It was very touching and impressively humbling! We Americans are so very lucky and fortunate.

SMSgt. Allred was the NCOIC (noncommission officer in charge) of this huge undertaking and had developed the personnel roster and purchased all the tools and parts necessary to complete the electrical portion of the project and help develop a plan with engineers and other subject matter experts involved with the project—totally site unseen. He also organized military airlift and housing for the trip down to South America too. It was an amazing effort on everyone's part, especially Harry! The project was a total success.

I had tried Harry's patience regularly during the first five years in his military shop, but through his unique method of encouragement, I had earned his respect. On this deployment, SMSgt. Allred gave me loads of responsibility, and I did not let him down. I will say this though, I had to come to terms with a lot of my personal issues, and I promise I am still working on some!

Our unit and support personnel from other units spent about six months working on this project, and finally several units from the army followed to help complete the multifaceted undertaking. I've often wondered how I became so lucky to serve under someone so skilled. SMSgt. Allred's influence helped countless technicians with their readiness, which in turn helped the countless people we served. I will go into some more detail about my Air National Guard adventures a little later.

After returning to Salt Lake from Boise and getting into the Utah Air National Guard, I had—through a high school friend, Rory—gotten a job at an automotive shop called Four-Wheel Drive Engineering. I worked at the shop during the week and at the Air Guard Base on the weekends. Most of the employees were from my high school, including one of the owners and project engineer, Randy, who had been Kearns High School's first senior class president! I found it to be pretty comfortable getting a job with old school chums. Randy was the boss but was also a mentor to us younger guys; he was a mechanic, and at times, a father figure. The other owners, Ron and Chico Aste and Bryce Bishop, were great to work for too and encouraged us to work hard and buy stock in the company! They gave us Christmas bonuses and treated us with respect.

Four-Wheel Drive Engineering built the very first four-wheel drive minitrucks for the major auto manufacturers, including General Motors, Ford, Chrysler, Toyota, Nissan, and Mazda corporations! We also converted vans, fire trucks, ambulances, and school buses to four-wheel drive too! The mechanics worked as a team on many of the conversion processes, and we each had our own specialty that we seemed to have naturally gravitated into and became quite proficient at. Rory was our best welder and taught the rest of us how to get better at this craft; he was also our exhaust pipe expert and was a master at it! After Rory taught me how to properly MIG (metal, inert gas) weld, I eventually became the shifter mechanism engineer and installer. I liked designing shifter mechanisms! Being an artist and perfectionist by nature, my goal with shifters was to create the most aesthetic and smooth-operating shifters possible. At the shop, it was totally possible to excel and gain true ownership and have pride and not get into trouble because of it! However, one morning, I was bending a three-quarter-inch solid cold rolled steel shifter handle into its final shape when something went wrong. The hydraulic bending machines (which operates at a relatively high pressure) forming jig slipped out of place. Instantly, it spit out the shifter handle—striking me lengthwise and somewhat diagonally across the right side of my face, crushing my right nostril, and fracturing my right brow! It had also torn my right lower eyelid almost all the way off and punched a zerk (grease fitting) right into my eyeball and rounded things out by giving me a concussion! Randy rushed me to a hospital, the medical staff there got me put back together, and I made a full recovery, but getting back on the horse for me was a little scary. Obviously, the pipe-bending machine could be dangerous, so I built a shield for the jig that got me hurt. It's worth noting, however, that I still crouched nice and low when I used that machine afterward!

I mentioned somewhere that I am an artist! When I was seven years old, I became a member of the Fisher Body Craftsmen's Guild. This is an organization that designs the cars of tomorrow yesterday! The Craftsmen's Guild inspires and coaches young boys and girls to become designers of the future! I created a model of a futuristic car way back then but never competed for a chance to become sponsored into the program. I have never forgotten the experience though, and seeing other kids' designs from way back then that are on the road today is inspiring! I had dreams like the other kids, but theirs moved on to become visions and realities with car design. I eventually got to see many of my dreams become realities too while working at Four-Wheel Drive Engineering! Think about it, a small penny stock company built the first four-wheel drive minitrucks. I feel quite proud knowing that I was part of the creative circle relative to the Toyota Tacoma that I drive today! All of this is despite the fact that I came back to Salt Lake disillusioned and angry about my childhood! Randy, like Harry, gave me a chance to succeed and showed me the way!

Around 1979, the auto manufacturers started tooling up to build their own 4WD minitrucks and vans. The writing was on the wall; our sixteen-man operation, including the secretary's job, was in jeopardy. We would still build fire trucks, school busses, and ambulances; but the minitruck business is what drove our company's cash flow! Our business was getting slow and the pay unpredictable, and I had just gotten married to the girl I had danced with in junior high. Remember, I stepped all over her feet! Teri had two little girls, Angela (three) and Melanie (almost zero), she was six weeks old when Teri and I started dating! Things were tight money-wise for us, and lately there were times when we couldn't cash our paycheck. My dad had remained in contact with me and offered to let me sell his chemicals on a part-time zero-commitment basis. So at lunchtime, while the other guys played basketball and had lunch, I would spend my time walking up and down the third west business district near our shop, selling hand cleaners, degreasers, paper towels, and anything else folks needed! Part-time chemical sales was pretty profitable for me, as there were many types of businesses in this section of town, it added to my Air Guard drill weekend pay and filled in some of the holes my full-time job at the shop was starting to create. I'll say one thing for my wife, Teri; she is the bargain hunter from hell! By using coupons and sniffin' out deals, getting lower interest rates, etc., she has helped our family have a much better life than I could ever have done with my income and/or hers. Teri is where our kids get their smarts from, period! I really had a good time working at the shop though, I learned so much about building and taking care of cars, and even had become a shareholder. It was a fun job too; while at work, the shop guys would instigate grommet wars, which really ticked old Randy off! Grommets are rubber bushings a couple of inches in diameter with a three-quarter hole in the center and are used to mount shock absorbers on vehicles and are fun to throw at each other! We saved the old ones when we installed new shocks on our trucks that we were converting to four-wheel drive. When Randy (our taskmaster) would leave the building, it was common knowledge that a grommet war was sure to happen. We'd choose up sides, which were usually decided by where you happen to be working in the shop, then we'd go at it. It was fun as hell and was a great stress reliever! Most grommet wars ended as a result of Eddy "Mountain Man" Arco getting hit in the eye or Paula, our secretary, coming out to yell at us for having another senseless (to her) grommet war and would inadvertently get hit in one of her breasts. Then she would get real mad and threaten to tell Randy! She would start cluckin', "I'll tell Randy, I'll tell Randy!" On occasion though, Randy would catch us in the heat of battle, chastising us with, "Every time I leave, a grommet war happens, how is this possible, it's like we have nothing else to do!" One time we were chucking snowballs in the shop when Randy walked in with a customer! Presently they were standing abreast of me! Well, a snowball had already been launched in my direction and nailed the customer right between his bespeckled eyes, exploding like a Fourth of July starburst rocket. His head rocked back and his jaw went a-slack from fright, and both his lenses were completely covered in snow. We thought it was hilarious, but Randy and the victim saw it differently. Man, it was the late seventies, and we raised hell on this job, built cutting-edge trucks, and, occasionally, after work, had great parties! We all went boating, fishing, and hunting together and even went broke together, and this was a place where I fit in.

I had also been interested in martial arts since I was about fifteen. Around the time I moved back to Salt Lake, I began studying under Rick Hobusch and Toshio Osaka, whom I had met through my brother Jess. Toshio was the local master of the Japan International Karate Do center and also taught karate at the University of Utah. Toshio had been sent from Tokyo to teach Wadō-ryū karate in the United States! Most of the guys at the shop were interested in karate to one degree or another. We often practiced at lunchtime and would hide between trucks and attack one another throughout the day testing our awareness skills and our boss's patience.

Dennis Vandenberg (our head mechanic), Rory, and myself were the most serious practitioners. Dennis and I studied with Rick and Toshio regularly (before I got married) and really enjoyed the confidence and physical fitness levels we obtained. This style of karate exposed us to real-world core values as well (e.g., true devotion to others and self), we learned to face our problems head-on and to be flexible, we also learned to be humble and how to make decisions on the run!

One night while on a date, while my girlfriend and I were walking down an alley, we were confronted by a group of four bullies who had surrounded us! We were trapped against a stonewall of a building, so I pushed my date behind me and told her, "When I say go, you run to the street yelling for help." The apparent leader (who was to my right) suddenly said, "It's a great night to get mugged, ain't it?" My defense system was totally on, and my ki (internal energy) was flowing like an A-bomb! These idiots were already toast but didn't know it! As the leader approached me, my plan was to take him out with an overpowered front thrust kick (he wouldn't have gotten back up) and then take out the next biggest threat to my far left! That done, I think the confrontation would dissolve. I projected my will into the leader who must have sensed the approaching war and backed down whishing us a good evening (I was relieved). If he would have come one more inch, my radar would have started the brutal and unforgiving attack. Had I taken action, life for everyone involved might have been for eons very different! Afterward, despite my controlled intensity, I began shaking and felt like throwing up. Toshio taught us to never fight unless it was absolutely necessary! He would say, "Use your wit, keep your distance, then if you have to, protect yourself, go all out. Take no prisoners."

Randy was the opposite of my dad. He was gentle, understanding, and a teacher, but he was also supremely determined to succeed, as was my dad! I had made mention earlier that Randy cosigned with me for my motorcycle that was sitting up in Boise waiting to be sold. Randy also loaned me his personal truck in midwinter to go up and get the bike—whew, what trust! He did say, however, that cosigning for me meant that I had to stick around for a while and that he couldn't fire me either until the bike was paid for. Randy owned an old 1948 Willy's Jeep but wanted to build a replica of it with newer or brand-new parts. He bought a 1972 CJ-3 frame (I think, could've been a CJ-2) and a stainless steel body with the stripes polished in. He bought my brother Jesse's '65 Impala for its high-performance 327 and obtained a Muncie transmission; now we had a project! The jeep took us six months to complete in our spare time, but when that thing was done, wow! The bumpers were a quarter-inch-thick, stainless with the taillights cut in. It had a tilt steering wheel, captain swivel seats, and a digital dashboard too; now remember, this was around 1979. The Willy's also had a custom grill guard that we made, a Warn winch, custom roll bar, black soft top, and BFGoodrich AT (all-terrain) tires mounted on spoke aluminum rims, and of course, a custom (by Rory) exhaust system! This cool jeep is a one-of-a-kind and has even won a few car shows at the local off-road shops, and that Willy's is still going strong today! Randy had taught us many good and useful things—to include compassion and survival.

Just a thought: Meeting these people—that is, Harry, Randy, Toshio, and Rick—gave me lots to think about. I was being given the chance to see that I had value and was intelligent. I realized that if I used my head, I could find a path to success and maybe even influence others to do the same. Their combined lessons have guided me in life to this day, but that doesn't mean I have always followed them like I should. I see myself as a work in progress, which means some of my lessons will be learned the hard way—through cold, hard, hands-on experience! Sometimes this is necessary, but there are often better and simpler alternatives. I leave it to you to choose your path.

Remember, I had a lot of stored and saved negative beliefs or ideas that needed changing and erasing, and these can take a while to fix. Which by the way, in my case, would involve many life-changing painful experiences to occur that would have an effect on not only me but those within my sphere of influence.

The question becomes "how do I find balance given my history thus far?"

6

Evolve and Learn, Adapt or Else

Around 1979, I had to quit the shop. I needed to start making consistent money, for me and my family, so I started selling chemicals full-time. Randy said he could never fault someone for trying to better their self, but I could tell he was unsure about the choices I was making. Randy believed the shop would manage; after all, the shop is still in business today! My dad had been talking to me for a while about working for him full-time selling commercial and industrial chemicals and other wares, it did seem appealing, but from the word go, my wife was against the idea. Direct sales is a tough business to be in, especially if you are paid only on a commission basis, which means "no sales no money!" All I knew is that I was going broke. My dad set me up on a draw where he would pay me $750 dollars at the beginning of the month, but I had to pay it back at the end of that month, he also gave me one day of field training and he didn't play favorites. I had to establish a customer base that included about two hundred regular buyers and a dozen or so frequent buyers that would help fill in the gaps of varying regular sales. My territory involved the whole state of Utah, plus I could venture over to Colorado and Nevada if I felt industrious. Lastly, I would have to buy my own health insurance. Being a salesmen requires you to be (if you want to be successful) very timely, a good listener, highly organized, honest, and sincere! I found that at first, I was so worried about making money that I never really spent much quality time with my folks. My dad, who was now my boss, asked me one time just how many potential customers I was seeing a day, I said forty or more! The lesson here is—this was why my sales were in the toilet, I was seeing too many. Also, there was a bit of cutthroat tactics going on in those days, which hurt the honest and ugly guy's sales. One hand soap company in particular would attempt to steal your business by giving your client the pricey chrome-soap dispenser for free and at least the first case of hand cleaner for cost. However, this outfits hand cleaner was forty bucks more a case, and in the end, the customer got bit. Yes, it was a real good hand cleaner, but there were about three to four that worked really well and for forty bucks less a case! Gratuities were another tactic used to buy business back in those days (e.g., TV sets, vacations), even under-the-table cash along with pens, screwdrivers, quality knives, etc. were shelled out to get and keep business.

One day, I went into a local major brand auto dealer (it's all we had back then) to sell some soap, sealants, and degreasers. My customer told me that I needed to split my commissions with him or it was no dice. Even though I was pretty young, I gave that guy a ration of crap that he didn't like too much. So he threw me out of his shop and said, "Good luck selling anything, dumb ass!" I called my dad and told him what had happened, which made him real mad; he then told me to keep my chin up because there are plenty of honest people out there too! He told me to remember also that there is no job that is below a man's dignity so long as it is honest! Later that week, I went to visit a customer named Kevin at a local steel pipe supplier. He told me that a soap company had been in to see him and had promised him the moon if he would change to their product. Kevin recognized the liar mouth's agenda and told him to go somewhere else to be dishonest! Kevin's action made me feel pretty good and proved my dad's point! The reference to the ugly salesman is because of a company that hired gorgeous women dressed in a way that would entice buyers to say yes by using the phony come-on. After the hook was set, some victims probably said, "I'll buy all the soap you have and whatever else you have." Man, that was tough to compete against. I couldn't even give my products away in the face of that competition. Luckily, their high prices got old as did the sales pitch!

One time I went into a Porsche dealer to make first contact with their buyer. He liked my product, appreciated my sincerity and promise to take good care of him and his needs. However, he told me that every single salesman he has dealt with said the same thing and never came back. "So we'll see how you do," he challenged. I never went back! Crap, for whatever reason, I joined the ranks of a bunch of guys (maybe even those hot girls) that were considered to be unreliable, and I have never forgotten it either! My sales were really never very good, but my dad told me that I was learning the ropes to surviving this business and I was "turning the corner." All I had to do was stay tough for a bit longer! But I had a new son, Eric, in the house we couldn't afford, our credit card was maxed, and with other bills mounting and cars breaking down, Teri and I were at odds and very stressed.

One Friday, I was driving home from southern Utah on bald tires, bad brakes, and was low on gas. I pulled into our driveway and decided to go to the mailbox to check out the next wave of bad news. There was an envelope from Publishers Clearing House that said "You Are a Winner!" The window on the envelope showed a check with lots of zeros! I started hooting and hollering, "This is it! I'm rich. I'm quittin' my stinking sales job!" I was making so much noise that my neighbor Wayne, who was a painter and had given me work from time to time, came out to see what the commotion was all about. Well, when I opened that envelope, I realized I was in fact a winner of ten dollars, which bought me a haircut and enough gas for the car so I could go to my drill weekend and earn fifty bucks. Back then, I was too proud to call my Guard supervisor for a lift, and I would have given myself the haircut! The ten bucks was something though, but it's like Wayne told me, "You are a winner," and I had proof.

Realistically speaking, soon afterward, I had to call my dad/boss or boss/dad and tell him that I was packing it in; my wife and I had decided that I needed to get a regular job. My dad pleaded with me to stick it out a little longer, that I had established a growing customer base and things would improve! These days, I know he was right, as many of my customers are, decades later, still buying from his company and they ask how I am doing. But right then, things were too tight; we wound up losing the house and moved in for a few months with my wife's parents, who lived in Kearns, Utah, which in itself was a stressful situation for everyone! My father–in-law, Howard, an old Navy guy who became a prison guard superintendent, seemed to be the most understanding. He was a quiet, hardworking man and honest to a tee, paid attention, and understood the struggle.

After a bit of morphing, I eventually got a job with the Hilton Hotel chain as a security guard, and we were able to move out of my in-laws' home and rented a duplex nearby. I also was getting work through Harry out at the base, which helped too. Of course, our country was right in the middle of the recession, which hurt lots of people financially. Amazingly, I had electrical friends complaining about being out of work but refused to leave their trade and/or refused to make less than twenty dollars an hour, despite their wives slaving at some minimum-wage job. They thought that I should apprentice with their companies and join them on unemployment, that it was undignified to be a security guard, but I needed to work!

Being a security guard for the Hilton was interesting. I worked for a Salt Lake city detective Wilf, who was moonlighting at the Hilton! Wilf, who liked to call me Olly, thought that I had a knack for detective work and was steering me away from dangerous electrical work and into police work, specifically detective disciplines. Along the way, I worked with the vice squad and went undercover several times to break up drug deals, which could be pretty dangerous! I also wound up arresting petty thieves, drunks, unruly vagrants, and a few hookers. To me, the bad part of being a cop of any sort is all the reports you have to write and the ridiculous court room dramas that you must endure! Seriously, I was in court with a robber that I had caught red-handed and arrested for stealing Hilton Hotel TVs; I was made to feel like the bad guy. The judge and I were quickly becoming "nonfriends" as this freak and his lawyer wormed this criminal a plea bargain deal. I could not believe what an officer of the law must put up with! I made quite a few remarks, one of which was directed at the caped crusader with the wooden hammer (not Maxwell's, his hammer was silver), and as a result, a contempt threat became directed at me. Wilf grabbed me, saying, "Hey, if you're gonna be a cop, Harry, you are going to have to be more polite!" That event decided it for me, there was no way I was gonna be a cop knowing how things are. So I quit this job and moved on to building maintenance with a local school district and maintained my membership with the Air National Guard.

My guard unit had taught us some disciplines that were not all that dissimilar from the ones Toshio and Rick exposed us to or the ones the cops live by. Most of the military sergeants I came into contact with had a lot in common with Toshio and Wilf; primarily it was about surviving! As our unit traveled around the globe, the new bees in our shop learned electrical disciplines from the old bees. We learned how to learn and to keep learning and how to adapt, which is a very important thing in any walk of life. There were many electricians in our shop, about twenty-two at that time, with a large diversity of skill sets! The skilled guys helped tutor us apprentices in our specialty known in the Air Force as an AFSC! Nick was one of our linemen (the monkeys who climb poles) and was a real good one. Nick's pastime back then was rooster fighting, which took place on Saturday nights! As such on Saturday drill, Nick would have his roosters covered up in gunnysacks in the back of his pickup truck in our unit parking lot. I would wait till Nick was out of sight, then go and pull one of the rooster's feathers that would be sticking out of a hole in these holey old musty-ass gunnysacks. It would get those rooters to crow like crazy and drive Nick nuts! He could never figure out what sent his roosters into those crowing frenzies, and maybe they even woke up a few people.

At some point during the midseventies, Nick and I had been assigned to disconnect a service drop to a building. A service drop is the wires you see (called triplex) crossing through the air from a transformer on a telephone pole to a business or residential structure bringing you electricity. We got our bucket truck located and boomed up to the transformer, and then Nick asked me to cut the secondary/low-voltage conductors, which is an expedient methodology, carefully done with a set of cable cutters. I chose instead to do it with set of eight-inch diagonal "side cut" pliers. The transformer was still energized, and a greater margin of safety needed to be in place! Because the tool was small for the task, I had to bite through several strands of wire at a time and had to really squeeze hard to make the cut. Nick was busy with something else and should have checked on his apprentice more closely, but I am sure he thought I would use the proper tool. As I began to cut the final strands of the last conductor, the effort it took seemed to break them more than cut! This action caused me to fall forward into one of the other secondary lugs creating a big arc flash. The subsequent explosion blew me back against the bucket we were standing in then recoiling me back into the transformer, causing a second blast to occur. Nick was freaking out, yelling, "Oliver, Oliver, are you okay?" When the smoke cleared, my face and head had been mostly protected by my hard hat, my arms had been protected by my work jacket, and the lower part of my body was protected by the bucket itself. Luckily, Nick was unhurt too, but was badly shaken and really pissed off, and after checking me and him out, we lowered down and went and got a cup of coffee. Then I got a much-needed lecture about using the right tools and asking questions when unsure. Nick flatly stated, "Ya know, Oliver, once you fire the bullet, you can't bring it back." Man, we were both so lucky to be unhurt! Accidents happen all the time, and in every nook and cranny in life, the causes are many, some simple, some complicated, and most of them avoidable.

Now, for a moment, I would like to talk about using the right tools for the job! This of course applies to the obvious, which could be hand tools etc. However, there is another set of tools that need to be used as well, they are as follows: communication, knowledge, and skill, along with good old common sense and common courtesy. These particular tools should be adopted by the leaders and the followers, and I define this action as a drop-dead requirement, period. However, the common sense part may be a toughie! Because Nick and I were not communicating enough, and I did not have a clear understanding of the proper process required to complete the task in its state, a potentially life-threatening event occurred. Typically, an Air Force electrical shop holds a "tailgate party" before it begins any job! The goal is to go over the details of the impending task(s) and to make sure everyone is clear about their responsibilities and limitations! Some jobs are so routine, however, that these briefings are not utilized, and usually everything goes well—usually. In mountain climbing, we have a saying that goes like this: "Your team is only as strong as its weakest member." I have found that adage to be very accurate and a good thing to remember! Nick's and my job was routine, but we were also working it "energized," adding an element of danger to our task; and because of me, it got out of control for a moment! A moment of time in regard to electrical current running loose can become a lifetime of sorrow or just as easy as "no tomorrow"! As we go along, we will talk some more about some other incidences I have experienced or became aware of in my career, but for now, let's shift gears and talk about the abuse side of the fence a little.

My viewpoint is that each individual responds to the various levels of treatment in their own unique and paradigmatic way. I have worked in a handful of abusive environments over the years, and the abuse was usually coming from a maladjusted leader of some sort! There are times when you don't spare the rod, but most of the time, reasonable treatment gets better results and may create learning and growing opportunities instead of bad feelings all around! I've seen good employees so stressed and worried because of a tyrant type of boss that I thought they would commit suicide! Most folks don't mind being talked to, but *down to* is so demeaning! It is possible to get a point across without making someone feel stupid or threatened! Also, there are leaders who create cliques with some of their employees, and those cliques seldom get very dirty. Recently, I worked for some people who were very threatening and found myself well out of the clique and in fear for my job, and I can tell you the stress was deadly, life-ruining, and cold! I felt like I was repeating my childhood, and I really didn't want to live in that form of prison again! Some of my coworkers and I checked on each other often to make sure that something stupid wasn't on the menu, it was that bad. I really hate those so-called leaders for what they did to us! It was a terrible thing for a group to have to live with, and it lasted about two years until higher powers intervened!

Speaking of how people react to adversity, Stewart Bates—clearly one of the funniest, nicest, most caring people ever—was badly injured when he was a kid. One day when he was about two years old, his mom heard him cry out! The best they can figure is that he tripped and, as he fell, pushed the corner of a bed frame into his into his right eye, permanently damaging the cornea and blinding him. Another time when he was five, he was hit by a car, severely damaging his head and legs, requiring hundreds of stitches and medical care to repair this little guy who was very lucky to be alive. How Stu maintained his loving and outgoing nature throughout his life is truly amazing, he just worked through and adjusted to the issues and had faith that everything would work out. Stu and I worked for Granite School District throughout the 1980s at a local high school in a suburb known as Granger in Salt Lake City. I was the afternoon shift maintenance manager. Stu managed our swimming pool, which was in a separate building on the premises; both of our shifts ended at midnight. One instance, around 1:30 a.m., well, after I had secured the premises and set the alarm and gone home, I got a call! It was Granite Police, they said that my building was unsecure; they claimed they had found sixteen front doors unlocked, I was in shock and knew that I would be fired! Every night maintenance manager that got fired was let go for a similar reason and usually on the first offense, just like the guy I replaced, Old Don. Really, there was no way I left sixteen front doors unlocked, it was not possible with the secure and lockup routine that you followed. Don had left one door open; knowing him, I could not believe he missed the door they claimed he did.

Security told me to come back to my facility and resecure it and that a Granite Police officer was waiting there to greet me! Great, so I called Stu, who graciously offered to come out and verify as I resecured the property while the cop stood by and condescendingly watched the one and a half–hour process. In the end, I got lucky and was given a second chance due to my convincing testimony that created reasonable doubt in my favor, that is, in the eyes of my boss, the principal! Stu's presence certainly helped my blood pressure and illustrates his devotion to friends in need! Stewart eventually moved on to higher callings in the real estate world! But the day before he left, Stu brought his dad to my office to meet me; immediately I chimed, "Well, if it isn't Mister and Master Bates." Stu thought that remark was pretty funny for a guy with a slow leak in his head and laughed, but not his dad, who frowned! Months later, after Stu had left his employ with the school district, on a Sunday, Stewart and his family came by my house to visit. Teri meanwhile had something very tasty-smelling cooking in the Crock-Pot, the aroma filled the house. I eventually realized that Stewart and his family seemed to be sticking around. My instincts told me something was wrong and to invite them to dinner, but I never did! Stewart after a while commented that they should go. After he and his family left, Teri and I realized that they were hungry but was too humble to ask to stay for dinner. Crap, crap, man—I went after them but could not find them, cell phones were not something people packed around in those days, and they were gone. I felt so terrible about delaying an invite to them. Why didn't I just ask my friends to stay for dinner? It would have made the afternoon more enjoyable anyway! My proud friend and his family left my home hungry.

Later I had the opportunity to talk with Stewart about that day, I apologized as deeply as possible, he put his hands on my shoulders and said it was okay, and we embraced. It wasn't okay to me; I failed a friend in need. I felt worse than ever because my damned instincts were right! Please never forget what I just said. Put slightly another way: "Your instincts usually are right."

To this day, I am so sorry for not having Stewart and his family stay and eat with us. The Bottom line is this: "Our decisions and actions will part the sea."

Communication, communication, communication. Don't we owe it to one another and ourselves to look out for one another? And it just might begin with educating to the best of our abilities, ourselves, and our followers at home, work, or play.

In the military, we are taught to train our replacements; this process maximizes survivability! I believe in training wide and deep to support this idea! What this means is that when a new recruit enters our shop, he begins OJT, or on-the-job training, which in the long term is preparing him/her to successfully do the tasks and run the shop! Along with the hands-on electrical training, we would teach our recruits how to manage their own training records and how to plan and schedule jobs etc. Everyone got the same opportunity to grow, and everyone needs to be able to step in at any level within the shop and make things happen; that was our mentality! At the end of drill weekend, the shop training manager, TSgt. Rick Bullock, would sit us down and have everyone make entries into their own 623 OJT record regarding M-16 qualifying, self-aid buddy care (an electrical job they may have done or a deployment that they just returned from etc.). The supervisors also would sign off portions of training that a person had completed, allowing the individual to move on to other training opportunities! The 623 OJT record was your documented proof of your military history—including your rank, SSN (Social Security number), and the career field you were involved with; this document traveled with you (or was supposed to) on all your deployments. Without it, you had problems! Our unit, I believe, exemplified the use of the buddy system just like the Boy Scouts! It's all about team effort and esprit de corps! We must live and breathe our core values: everyone counts, and everyone goes home alive!

I don't think I was a very skilled people manager while working for the school district; I simply wasn't mature enough to have this position! I did great at getting the job done but definitely needed to be more effective and fair with people; my boss at the time was very busy trying to teach me this lesson! One day he told me that the best thing to do is accept your staff and try to help them grow and work with them, have lunch with them, and just be part of them. Wow, what great advice from a guy who had been a union president for General Motors! I think that my military do-or-die attitude at the time got in the way in this environment that required more reasonable approaches to succeeding! I eventually learned this balance lesson as I made my way through life managing many other folks, including my own children! I just wish I could have learned it a bit sooner! There are lots of versions and practitioners of the authoritarian leadership mentality; later in my career, I remember being taught how improper it was! Remember, I went through boot camp in the midseventies, so life in the military and life in general then was still about strict disciplines; all we knew was that the mission must be a success! I still believe in mission success, but I have learned to be more flexible in my approach, which seems to increase my options and success rate to boot while still maintaining core values! If your people succeed, you succeed. Several of these older mentalities, to this day, are alive and well, especially in the workplace! One of these mentalities causes the, let's say, supervisor to attempt to keep their people uninformed, keeping need to know information out of the channels, and slowing down or eliminating necessary training etc. It's a mind game-and-control tactic. Some leaders cause so much stress in the workplace (so do some employees) that they create a dangerous environment! People become forgetful and get in too big of a hurry etc., but the reaper will catch them eventually, just ask any ER nurse. I have worked for people who were insulted and threatened because I was a highly and broadly skilled electrician. To them it was a problem that a subordinate was more capable on a particular subject than they were! To me, at this point, the leader is nothing more than a mistitled follower. Unfortunately, their subordinates are kept in the dark, held back, and made to feel as if they have little value and are easily replaceable and are berated during performance reviews etc. Amazingly, I have found that with many of these same subordinates, loyalty runs straight and true and will for a lifetime! Sounds a bit like the Stockholm syndrome, doesn't it? I know this is not true of all businesses, but it is commonplace when working for the government, where apathy is the name of the game. In the electrical world, learning its dangers—like arc flash and shock—the hard way is a great teacher if you survive it physically and/or mentally!

For fun and for example, lets break an arc flash scenario down, which by a loose definition is when there is a fault in the circuit caused by an energized wire bypassing its load and grounding out or going to a neutral or another energized conductor creating an arc flash. Okay, so you are working in a live 208/120-volt panel, and your metal screwdriver slips, and you short it across two bus bars. The voltage of the fault is 208 volts, but the fault current level (measured in amps) is rising exponentially as the alternating current cycles are happening! We are hoping right now that an upstream fuse or circuit breaker is going to trip instantly before several AC cycles have occurred. Remember, our electrical systems run on 60 AC cycles per second. However, the explosion or arc flash right now is already very intense and is a developing 50,000-degree fireball of molten metal that you are becoming engulfed in. In reality, you have already sustained some form of damage! In lay terms, what you did was get shot with a 12-gauge shotgun full of fiery pellets at point-blank range. Sounds exciting, huh? Unfortunately, you probably weren't wearing any arc flash protection, which has become mandatory only in the last decade.

These electrical dangers seem like they are something new, yet people have been getting hurt and killed by electricity for ages. Simply put, nobody paid attention to the dilemma until recently. Even now, someone may not give a crap about your safety enough to help you electricians and HVAC (heating, ventilation, and air-conditioning) folks understand the true ferocious nature of your job and supply you with the necessary training and equipment. If you are caught in an arc flash, you are now either dead or are a third-degree victim up in some burn unit and probably wish you were dead! From your grave or your burn unit bed, could you see the shocked, sorry, and decaying looks on the faces of your loved ones, including the doctors and nurses? Could you see all the financial troubles coming down the pike, the shattered dreams, the shattered lives? Did you notice the ripples from that pebble thrown into the pond? Do you now, do you?

I lived in a burn unit (I considered it to be a prison) several decades ago as a third-degree burn victim. One day, a new burn victim showed up bandaged from head to foot! I went over to his bedside to meet him and ask what happened. His response was that he used to be an electrician. I was stunned and told him that I was a "sparky," then asked him if he felt up to talking about the details of his incident (talking is good medicine). Struggling, he said, "Sure," then asked, "do you remember the rule about standing off to the side when closing (turning on) a circuit breaker?" I nodded, saying that I did. He then continued with, "I didn't do it with this 200-amp three-phase main breaker at a newly built commercial facility." Something unseen was wrong downstream from this breaker (*downstream* is electrical jargon for circuits fed by this breaker). Thus, when he energized the panel, which didn't have any of its covers in place, the breaker exploded, burning the front of his body to the third-degree level. You see, electrical circuit breakers have several ratings, the one most people care about is the trip rating (i.e., 200 amps in the case of the exploding breaker!). However, there is a second rating known these days as the AIR rating, which can be tens of thousands of amps of short-term withstand capability! If you are around when a 200-amp breaker trips on its trip rating, it may be impressive, but you will usually not be harmed! However, if you are too close when the AIR rating is exceeded (like my friend); on any breaker, potentially, you will be hurt or killed! Simple steps were not followed, it was a decision, then life went 180 degrees, you suffer, and/or you die! And don't forget the other circle of people involved—his wife, kids, dad and mom, and the coworkers are all affected too. Ripples, ripples, ripples. I don't know if he lived or if he even wanted to.

The burn unit was full of similar tragedies (e.g., there was a guy missing an arm and a leg and had third-degree burns over most of his body)! He had been involved in a boiler explosion, another guy had been completely buried alive in hot asphalt and was still fighting for his life, and poor communication was the cause here! A large dump truck carrying yards of burning asphalt dumped its entire load on this man, oh my god! Also, there were three little kids in the burn unit too—a toddler, a three-year-old, and a five-year-old! They had been playing with their mother's cigarette lighter and started their bedroom on fire. Apparently they were not burned by the flames but were burned and melted by the intense heat. Their mother pushed in to the blaze with a mattress as a shield, rescuing her burning, screaming children! My wife would come to visit me and spend some of her time with those kids, I'm glad she did. She would play dolls with the five-year-old girl who every so often would cry out in pain; Teri would comfort her as best she could. The little girl was badly burned; her level of bandages told me her truth! Mostly though, this little thing adapted to her new world pretty good. Many adult burn victims seem to resist the change in their life, which complicated their new set of "life rules," which either slowed down or stopped their recovery. Interestingly the children here seemed to do better with their issues. However, as adults, us burn victims have been exposed to a bit more of life's various processes, including tons of negative crap. We have been rejected by one group or another—and for the stupidest of reasons, like money or the lack thereof, material wealth, our skills, our looks, even the cars we drive; and maybe we gained some toughness, and it helped us survive. But now we are burn victims who get stared at out of the corner of some passerby's eyes, adding to the reasons we just don't fit in! It's almost as if this injury, with some people, brought the individual's negative aspects closer to the surface to ally with the injury itself. The man from the dark side has many faces and tools to use against us, whether we are aware or not! In some, I saw it bring the positive aspects forward too.

The world is far from perfect, and accidents can and do happen, but it sure would be good to avoid some of them if we could!

All I know is this: we have to survive no matter what the circumstances may be.

7

Attitude and Altitude

In life, attitude plays a vital role in safety and survival, be it in the home, workplace, or the many adventures we get involved in. Even if a boss or spouse tells you to stick your head into a turning fan, it doesn't mean you have to, and if you do, this clearly places some of the responsibility for the outcome on your shoulders, it has to!

You have the right to say no or question the outcome of the action, whatever it may be. Throughout life, I have heard an old adage that goes like this: if it looks wrong or bad or dangerous, it probably is. Heed this advice as a real and logical warning and "take two days, hours, or minutes to think it over." Most of the time, you can come back and try "whatever" again, but "once you fire the bullet, there ain't no bringing it back."

In the climbing world, its history is littered with injury and death, and a major portion of the accidents occur during descents; however, quite a few do happen while moving up a mountain or across its glaciers. A climbing descent is characterized as climbing back down a mountain ridge or face and may involve down climbing, hiking, rappelling, and Jumaring.

Rappelling typically is an action where one creates an anchor—either mechanically with artificial anchors drilled or set into the rock or ice. It is also possible to use large rocks or trees to attach large slings (known as runners) to for the same purpose! Then you feed a climbing rope through the anchor and toss equal lengths of said rope down the precipice to where you want to go. With a descending device attached to your harness, the climbing rope is fed through it, then one uses a specific "brake and guide" hand technique to carefully slide down the rope to a stopping point.

Many climbers have died rappelling, the cause usually is the anchor somehow failed or the climber slid off the end of the rope. This activity could be one of the most dangerous things we climbers do since we are totally reliant on the integrity and management of our equipment, including the anchor we set and our ability to control our descent and the environment.

Rappelling contains many variables that when mixed with us *Homo sapiens* and our nature and the environment, can easily create a deadly combination. For example, one of my Air Guard buddies Perry told me that his son and a friend had gotten into the sport of rappelling. Perry wondered if I would meet him, his son, and the friend at a convenient climbing area known as Pete's Rock, which is an eighty-foot-high chunk of quartzite sticking out of the ground on Salt Lake City's Wasatch Front. Pete's Rock offers various technical challenges and has been climbed by many of the famous rock climbers at one time or another. Perry himself is a bit of a risk-taker (a.k.a. Scary Perry) but seemed concerned that his son and friend were in over their heads with this sport rappelling gig! Perry indicated that they also wanted to learn about setting anchors etc. I was a bit concerned and told Perry that it requires more than an afternoon to understand all the nuisances of proper anchor setup. I agreed to come and check out the boys and their process, but he needed to get them into a professionally taught course if they were going to last very long!

So one day after work, I met the trio up at Pete's Rock, and right away I had to put on the brakes. These kids with *intensity* written all over their face had been rappelling with bathrobe belts wrapped around their waist several times to create a swami belt, which has absolutely no known strength testing! I showed them my Black Diamond rock climbing harness and ATC rappel device with a locking carabineer; they were unimpressed and said that stuff cost a lot of money and that their gear has worked just fine because they are still alive!

Now, they wanted to see how to set some anchors in the rock to make that part of their sport doable. So I told them that I would give them a healthy dose of advice, but they had to promise to get into a formal training process because if they didn't, they would die, and it would be because their system failed in part or in whole! So we set some big hexes (passive chocks) into a crack, equalized them and made sure the rope was protected from sharp edges! All the while I was in their ear about their gear, their knowledge, and their lives. These two cubs weren't all that interested in my sermon, and their attitude kinda made me mad. Before anybody went down our line (rope), I pointed over to the north end of Pete's where a popular climb and rappel is located. I told the trio that a kid was killed over there when his anchors failed. I encouraged them to feel free to go check out the big bloodstain where he cracked his head open, which was still obvious at the bottom of that route!

First to go down our rappel was the dad; I wanted to teach this parent and risk-taker a humility lesson. Perry knew what to do and seemed confident as he had been in a rappel class that I had recently given on a sixty-foot-high tower at an army confidence course. But now, he was eighty feet up an imposing rock face without big steel beams for anchors. Oh my, what a difference that makes in confidence. As he approached the vertical, he kept asking, "Is everything okay, do I look okay?" We all encouraged Perry, saying, "Lean back, buddy, and feed the rope as you walk down the rock, stop and look around a bit if you want, make it a nice trip!" His eyes were big as saucers, even though he had a ground-safety man. After he was done, he told me that he wasn't holding the rope back enough so he wasn't breaking easily and thus overgripped the rope to try and slow his descent; overall, this rappel scared him a little! I told him, "Think about what you just experienced," then reminded him again, "This is why your son and his friend need you to hold the rope at the bottom when they come down, just in case." Again, it is also why they can't be allowed to go out on their own without the proper gear and knowledge and the skill to use it. The day went by without incident, and I hoped that everybody learned something—I sure did.

When I'm climbing, I usually have to have several things go wrong to get me in Dutch! Please don't construe this to mean that the activity of climbing is less dangerous! If an anchor fails while I am climbing, it may not immediately be critical; however, if that happens while rappelling, usually you are falling, and your survivability is definitely in jeopardy, period.

Having been a climber most of my adult life, I can safely say that I have come across too many fledgling, climbers that never had a minute of qualified instruction, like our rappellers. Yes, I know that the forefathers of climbing invented our safety and climbing processes as they went, and used very questionable gear by today's standards. But they were definitely a different breed of people and, for sure, much tougher! But solid safety practices have to start somewhere despite the spirit of adventure or how tough you are. For those of you who don't like hearing all of this, too bad! Those guys (the forefathers) did a whole bunch more with a whole lot less, and they helped create the evolution of the great safety equipment and training we have today, so put that in your pipe and smoke it, ya big weenie. Anyway, now that I ain't mad anymore! Hypothetically, our little cubs above could have ventured out into "them thar hills" to become climbers too and gain some more false confidence! Yeah, yeah, I know, they might learn some good stuff too, but they probably would be clipping bolts (preplaced protection used in sport climbing) and getting good at moving over stone before they ever went into the real world to learn how to be a solid traditional leader where you gotta know how to route find, read weather, set them thar anchors as you move up the rock or ice.

Trad, or traditional, climbing leaders need to be mechanical experts and need to know when and how far to challenge their leading and climbing skills, or disaster could happen. Still hypothetically speaking, now our cubs are at risk because they never learned how to execute a solid safety system and have ventured onto traditional climbing territory. Things are going bad for them on a nice sunny day by pushing into realms way beyond their ability to cope; they may have learned how to move over some rock, but they didn't learn how to lead.

Self-assessment and being honest about your abilities can be something that not many people find to be a comfortable thing to expose! Sometimes you may think you are at the highest level of capability, and then you meet someone that is literally on a completely different level, and that can be hard to accept or admit. Having blind faith in humility is a great attribute to have in life.

I climbed with a chap named Shayne long ago! When we first met, we went through the formalities of comparing skills and techniques that we as individuals possessed as climbers. By all accounts, my man was far more skilled than I, and he said that he was a solid 5.11 leader on traditional routes as well as sport climbing; this also included all types of rock and environments! Obviously I would be his belay slave and be proud and happy to have the opportunity. For our first "climb" together, I suggested that we get acquainted on Bushwhack Crack, a moderate 5.8-rated rock climb on the beautiful granite of Little Cottonwood Canyon near Salt Lake City. This climb is a fun but demanding "crack" climb that requires good hand-and-foot jams to ascend and can be several pitches (rope lengths) long. Depending on your preference, you can link Bushwhack with a climb known as Mind Blow and another called the Hook, requiring friction and lie-backing skills, thus increasing the overall difficulty to 5.9!

All in all, the experience is awesome; all along the climb, you have a great view of the glacially carved Little Cottonwood Canyon and its dense forest complete with the sound of tumbling water from Little Cottonwood Creek far below.

When Shayne and I arrived at the base of the climb, we pulled our climbing gear racks out of our packs! Shayne's rack was scant; it contained seven or eight quick draws, which are roughly a ten-inch loop of nylon with two carabineers attached to connect to ropes and anchors. He also had several automatic camming devices that self-adjust in a crack to become a solid anchor when properly used. His rack may have contained a passive nut or two, which are similar to the hexes I used at Pete's Rock! Plus, he had a fairly new harness, climbing shoes, and a chalk bag, but was overall slim on gear. Incidentally, the older climbers from the 1960s balk at the use of chalk and call us new kids the powder puff boys. Anyway, Shayne was eyeing my rack with as much suspicion as I was eyeing his. My climbing rack was a pretty standard one for our local area! A trad, or traditional, rack usually contains a full set of cams and nuts (fall protection), which are doubled up on a few sizes. Typically you will also have about ten quick draws, several full-length (tied) runners, a nut tool (for removing anchors and pickin' teeth etc.), a harness, chalk bag, rock-climbing shoes, and of course, a 200 foot-10.5 mm climbing rope. Since Shayne would be leading this climb, I asked him if he wanted to take some of my gear. He replied condescendingly,

"I get leery of guys with too much gear." I was becoming a bit skeptical of my new partner! Climbing is dangerous, and there is no room for showboating, but maybe my friend was so good that even though he was on unfamiliar ground (he had never been here before), a light rack would be more than enough. We should have discussed it further. So we gear up, tie in, and I put him on belay.

"Belay on?" Shayne questioned.

"Belay on," I answered.

"Climbing," he announced.

"Climb on," I commanded.

The first move up Bushwhack has you stepping from a big flattopped boulder while mantling with one hand off a chicken head (a bulge sticking out of the rock) while you are jamming with the other and working your feet ever higher up the granite wall. Eventually you will stand on the large chicken head and will place your second piece of protection as high as you can reach. The first piece, sometimes placed lower (below the chicken head), is to keep the rope from working these higher pieces out of the crack! So piece number 2 is the first one that will hopefully keep you from decking if you fall out this smallish hand crack as you ascend above this location! Shayne really struggled to surmount the chicken head, but with some growling, he did make it. Of course, his belayer was dutifully encouraging his effort, which usually involves a bit of ribald comments; it really helps motivate your struggling buddy. Shayne then placed a cam just above his head and clipped into it and seemed relieved but also seemed a little worked and tense at the same time, totally understandable normally if you are climbing near your skill level.

Cams, by the way, require a bit of knowledge and skill to be placed well enough to do their job (all protection does)! After I told Shayne to check his cam one more time and suggested where to place the next one, he started up again struggling with the hand crack. I noticed that he was not getting very solid hand or toe jams, which are critical to this form of climbing. Shayne was slipping down through the crack as much as he was going up; this is painful and can wear you down fast! As he moved up, I noticed that his camming device—now about two arms' length or so below his feet, and the only piece of protection that would keep him from decking—was working its way out of the crack, and my friend's connection to the crack was shaky at best! I had to tell him that his lower piece wouldn't stop a fall and to get something in now! I know it scared him, but to let him get higher considering what was happening was no good either.

He immediately choked out, "I'm coming off!"

I fired back, "No freaking way, you jam those hands and feet in till you bleed, you ain't falling, damn it—do it!" I commanded.

As I recall, he quickly got in a second cam—which was questionable—clipped in, and he was done. I zipped him to the ground and checked him over, yep, he was bleeding, a bit scraped up and on an adrenaline overload, but he was otherwise okay. I hammered him about all those skills he claimed to have, he admitted that he had never climbed *granite* before. I told him, "You ain't no trad climber either." He nodded in agreement. After I was done ranting, Shayne asked if I was capable of retrieving his 150 bucks' worth of gear stuck up there in the rock. Calmer now, I was very glad he was not hurt but wondered if I should have not let him go beyond that first cam? I told my buddy, "You fought a good fight, and sure, I'll go get the gear," adding, "but only if you still wanna climb this thing." Shayne couldn't believe I would still climb with him! We ascended Bushwhack Crack and Mind Blows (5.9, face pitch, which involves friction, smearing, and micro-edging moves) and finally the Hooks "Flake," with its spectacular lie-back moves several hundred feet off the deck. After you have moved up twenty-five feet or so, you step right just as the Hook gets ready to hook ya left, out onto thin edges, then up another thin-edgy frictiony section; finally, it's on to the belay and call it a day! That afternoon, we both learned a lot, I am sure! We made two eighty-foot rappels and went and had a cold drink.

Accidents happen all the time, and there are so many variables that can and do interject into the situation that you may not be able to control, but then there are the ones you can control—some of which are knowing your own limits and being upfront about it.

In life and in nearly anything you do, if you do it long enough, you will have some sort of close call; there simply is no getting around it. In climbing and in all of its disciplines, we climbers will come face-to-face with this reality and must be prepared to deal with it. As an electrician, I know that anybody who is practicing the craft will eventually get shocked, and they may even get burned or killed, and there is no getting around that either! One simply has to know how to use the required safety equipment, put it to use, and maintain a clear focus on the task at hand. If you think for a minute that it can't happen to you or the people you climb with, work with, or manage, you are so wrong; it is why we must educate ourselves and use the safety mechanisms in place as much as possible! Tell me this: that since you have been driving, you haven't had at least one very scary moment in your automobile that you or someone else caused. I know you have, and you should feel lucky if you were wearing your seat belt, especially if you had a collision. Hopefully you also had EMS (emergency medical services) somewhat close by, and a passing motorist who new self-aid buddy care, and somebody who had a cell phone.

There is the perfect world, then there is the real world, then there is us.

Of course, you all know that I am the best climber the world's ever known, that is as long as I am standing on the ground.

8

Safety and Survival

Have you ever heard of first-, second-, and third-order effects? Basically, they are related to cause and effect; for example, a single misconceived decision can create a domino effect, potentially snowballing into a disaster. Over the years, I have found that safety classes, no matter how well prepared and dramatic they may be, generally only get the attention of the few, rarely the many! In the workplace for example, it seems that the managers and bosses are mostly in the "many" crowd. Managers count the beans and are interested primarily in the numbers related to how much work got done on any one day, and that is the bottom line, see! It seems to matter less that someone could be hurt or killed in the process, at least in some places.

Did I hit a nerve? Good! Maybe somebody has a conscience after all!

Several years ago, I was assigned to give a safety class to a group of government maintenance workers; I was allotted a whole two minutes for this class that was directed toward electricians and

HVAC folks. In my version of affirmative action, I started things off by showing a picture of a kid stealing wire from a high-voltage electrical yard in Louisiana. The kid was trying to take what he thought was any old piece of wire when in fact it was an energized conductor. After the electricity was done with this person, all that remained was a charred skeletal form! Our foreman (the "two-minute safety class" specialist) would only look at the photo out of the corner of his eye and swallow uncomfortably.

Our colonel exclaimed, "Hey, that's a body!"

Somebody else replied, "It *was*."

My two minutes was up, so I handed out a copy of the picture to everyone present, told them to be safe, and left!

Here at this workplace, the attitude is that we are wasting our time and money with workplace-safety classes—we are safe enough and do just fine.

We also had a similar issue with protective masks. Our concerned foreman claimed that we weren't going to spend money on protective masks because none of our buildings had anything in them to worry about. He insisted we have no chemicals, dangerous dust, or asbestos in any of our facilities, so we would use simple paper masks, and that was that! Interestingly, as of 2012, we were still removing asbestos from our older facilities, which included office spaces, battery rooms, paint bays, and auto shops! We also have underground vaults that need to be sniffed out for the presence of dangerous fumes. The workers routinely work with chemicals that are considered hazardous and require PPE (personal protection equipment) and training before use. The decision to negate safety because of money is easy for the boss, because he never gets exposed to the threats that his subordinates do. In fact, and I am sure this is old hat, he is never or rarely seen in those workplaces.

At this point in my electrical career, I have been involved in or have witnessed or have become privy to quite a few electrical incidences: arc flashes, electrical shocks, and electrocutions. I am not proud of this at all! However, it has been, for better or worse, part of the electrical trade side effects; or to say it another way, it is what happens along the way. Yes, it sounds—and is—stupid thinking. Electrical shocks, arc flashes, etc. are all too commonplace. As a result, the safety-rule folks are busy developing equipment (finally) to protect us from mishap, but we can do better as managers and employees to reduce the number of incidents that take place. Attitudes toward safety must change, however, or all the safety gear in the world will not make a difference. Personally, I am amazed that a sparky can come back to work and be an effective electrician after having been directly involved in a serious electrical accident, some do not! What about all the other types of accidents that happen, and what about those victims? I have personally been "in the line of and on fire," electrically speaking. I know my comrades wondered if I still was capable of being safe and effective, I wondered too, I would wonder about them also!

One day back in January of 2013, I was rewiring modular furniture in a large office space at a National Guard facility. While correcting a connection issue in one of the cubicles, the tenant (a young lady) asked me if I was the National Guard's electrician, I said I was. She then asked if I was the one who had been "struck by lightning." I again said, "It's me, in the flesh and back from the dead" (my standard comeback). Right then, this gal's normally happy, bright, smiling face turned to total gloom and sorrow and became unpretty! She said, "You have a very dangerous job," and I nodded in agreement. She then told me that her ex-husband had been killed a few weeks ago in southern Utah while doing electrical work.

I was stunned by her openness but asked, "Which electrical company did he work for?"

She said, "He's not an electrician and didn't work for an electrical outfit, but his boss forces the workers to do work they are not qualified to do to save money." (An old story.)

She then told me that he was working on a pump motor in a wet environment with the power on, and he got shocked, it was just two weeks before Christmas, now her little boy's father is dead! All I could do is comfort her, and damn those idiot bosses who have taken their toll on people by using scare tactics! Yeah, yeah, I know you have the right to walk away and say no, but try it and see what it gets you; even some safety inspectors turn a blind eye! That's right, even the safety offices that are there to prevent these things from happening can't be everywhere and can be levered against what is right! HROs (Human Resource Office), they just want everything to go smooth so they can get in their rounds of golf without any headaches to think about! To be fair, not all these offices are so dysfunctional, however, but many are, and I have had the opportunity to work with some of them.

Ethics in this country is on its way out, and computer games are top of the heap now, as is money for nothing. We don't want to know reality; everything has to be easy and free. Back in the eighties and nineties, some of my Air Guard coworkers and I were on a backup airplane crash cleanup crew! Debriefings for the cleanup crew were a typical and necessary part of the process in the aftermath. The training for these types of events makes it clear that not everyone can handle these so-called cleanups. This training made me wonder about people exposed to other types of trauma. Do they not need help? What goes on in that hollowed-out corner of their mind, who fixes the heart and the soul? Having lived in a burn unit and seeing all the devastation, it got to me. People—or what used to be people—were missing limbs, chests, even faces and even themselves. I'll never forget the sights, sounds, and the smells; the trauma was unrelenting. The psychiatric department in my opinion didn't do me much good, it was their approach! To me, I was a number to them, a case with an apparent set of obvious issues. My treatment was a little too ABC 123—they were not treating *me*! They didn't take the time to get to know me; I knew it, and it ticked me off, I'm still mad. As for other victims, some just fall off the end of the earth. I think careful understanding of and getting to know the individual and caring about their well-being is vital! People die from injuries, physically and/or emotionally, all the time. I can tell you that I personally was not very motivated to try and survive by living in the burn unit with people on death row! As a result, I did not see much light at the end of my tunnel. What I'm saying is, there is a mind-set out there that is very destructive and is omnipresent! Management spends way too much time not following the protocols set forth to ensure you have competency and safety necessary to the environment for your workplace and the people within it. Of course, there are accidents; that are just that, accidents.

However, we will hire people who have little or no skills just because it's a friend or their son or daughter etc., thus weakening the infrastructure as a whole! Some government agencies are famous for this practice! So now you have someone who might be in a technical career field who will need lots of babysitting, or they may get injured or hurt someone else! Also, people are hired who are "nobody's friend" but are still unqualified for the task, and they still represent a danger to themselves, the people around them, and the facilities they work on. We do not use the probationary periods designed to manage a person and help them grow or move them on to places where they may be more productive. If you are going to hire totally unqualified or under-skilled folks, you as a manager need to, out of fairness to everyone, at least get the "unsub" some qualified training so they can grow and become useful. After all, we have taxpayers who will fund the event, be it training or the probable workman's comp fallout! Yep, of course, and I'm sorry, the taxpayers get it in the shorts again. The boss man, on the other hand, is swearing everything is or will be just fine; after all, the sun is out, and the little birds are tweeting, and our drive in the country is so divine. This same boss makes other great errors too, which could cost everyone a pretty penny; he doesn't even bother to check if his crews are all back and accounted for at the end of the day—he just assumes! And if someone above him asks, he states with confidence, "Everything is fine, everything is just fine!"

Case in point, an HVAC worker was working or hiding in a crawl space during work hours and came into contact with an exposed electrical line and went into ventricular fibrillation! He lay there dead for several days before anyone thought to check on his whereabouts (Wow, I wish I worked for his boss, what a great guy!). Amazingly, the news said very little about this incident, it blew over fast, do you wonder why? Well, the suitors can't look bad, can they! That operation was typical—of all the similar operations I've been around, and the prevailing attitude is, "We work for the state, and we can't be fired!" Sadly, this is what is meant by "living off the lamb," when it really should be a badge of honor. Wow! This can of worms is a book unto itself and a big long disgusting one! On the flip side of the coin, accidents can happen to more-experienced folks too, which can be very costly physically, emotionally, and monetarily.

Years ago, Clark J., an electrical buddy of mine, and I was working in a live 208/120-volt 200-amp panel in a mechanical room at the Air Guard Base. Back in those days (the early nineties), we hadn't heard of nor seen insulated tools, much less arc flash gear; all we had were voltage-rated safety gloves, safety glasses, and guts! One afternoon, while working in the mechanical room of a sheet metal shop, we were tightening a branch circuit connection on an old general electric (GE) panel's neutral plate. Suddenly, my non-insulated straight-blade screwdriver slipped off the screw head and grounded against the back of the enclosure then shorted to one of the three live buses, creating an intense arc flash of about one and a half feet in diameter that shot straight out of the panel box.

Fortunately, only my left arm (which was covered by a long sleeve cotton shirt) was in the direct blast path, so I did not receive any damage there! However, the left side of my face was just at the edge of the fireball and did receive radiation burns, including my eye, nothing serious though! Clark and I were scared out of our wits, and Clark's straight-blade screwdriver was ruined. The other fortunate thing was the circuit breaker protecting that panel tripped immediately, shortening my exposure to the fire and the 50 thousand degrees Fahrenheit of flying molten metal! Unfortunately, the breaker was permanently damaged during the fault and had to be replaced! Thank God for OCPDs (overcurrent protection devices), a.k.a. circuit breakers or fuses.

FYI. Note: When there is a ground fault or a line-to-line fault, which is caused by a live wire touching metal casings or something grounded, or when two hot wires of different potential come together, the upstream breaker should trip. The trip rating might be 200 amps, as in our case. However, these sorts of faults, by nature, create large inrushes of electrical current that grow to heinous levels as the clock ticks away in microseconds and are invariably much larger than the trip rating of the circuit breaker. It only requires a few cycles of AC current (a few tenths of a second) for the current value to become exponential in volume, thus hideously more destructive.

This is why we hope the breakers or fuses were engineered correctly and do their job when they should; life is better for everything and everyone when they do.

Some of our apprentice electricians were present and witnessed the event that Clark and I created. One of them, Ed (a former Marine), was standing near a large compressor about a dozen feet from the panel that just impressed everyone. While Clark and I rectified our new problem, this large compressor turned on with all its might. Its belts started squealing bloody murder while the big motor raged, making lots of racket; it scared poor Ed so bad that he bolted out of the mechanical room, and he quit the trade. We wondered if Ed had been already very afraid of electricity, and this combination of things pushed him over the edge. Would his hidden fear have caught up to him someday, especially if he had to work in a live scenario, and this fear clouded his judgment? Clark and I were both affected by the fact that my slip scared him so badly. For Ed, quitting the trade was probably a good idea. I still feel bad for my slipup scaring him so badly, then to have the compressor finish him off!

Fear, however, is a beast, and it is as real as your mind thinks it is, even if there is no tangible reason to be afraid. When fear is in control, it causes all sorts of negative things to happen, like hesitation, which causes your fears to come true; then, the beast eats you. The experiences I have mentioned are unfortunately more commonplace in the electrical and HVAC trades than we would like to admit. They sometimes become reality due to impatience, laziness, lack of thorough training, being overzealous (because of great experience and knowledge or ego), time and money (there's that phrase again), being too tired; being pressured by leaders, safety inspectors, and even customers create the altered playing field rules! These are the excuses that we use for rules to play with our lives! Although I can say that being dead ain't so bad for the dead guy or gal because you now have no issues, no tires to rotate, no taxes to pay, and no getting stuck on the freeway! However, the ones that you left behind don't have it so grand, they still have to get up in the morning, go to work, and remember that you are gone for good; and for that, they will have to pay many types of costs. Even if you do survive some accident, but are too damaged to be useful, your prison and the one your loved ones exist in will really suck! Remember, there are many forms of pain—that is, the one(s) from your injury that are physical, and then the mental anguish. Then there is the pain in your friends and loved ones' hearts and souls. Yep, you and these people will get to experience all this; maybe even your boss will too! The reality is the laws of physics and averages and Murphy usually hold all the cards that matter. Hey, don't get the idea that this don't apply to you 'cause you ain't no sparky; it applies to truck drivers, office workers, carpenters, housewives, people at home and at play, even cops and robbers, and ministers! In some cases, natural selection will be the final rule. Man, that has a finality and scary ring to it, doesn't it? Seriously, I think my friend Ed made a very smart choice as he ran from that mechanical room long ago. I should have run with him.

I once was told a passage on safety that went like this: Out there in our endeavors, we take little and big risks over and over. In the beginning, we knew it was unsafe, but we kept taking our risks and getting away with it! Our rational eventually justified the act because we never got hurt, and what we were doing seemed even a bit safer than it was yesterday. Nowadays, this same risk has become so commonplace and okay that we teach it to others, even to our own children. We never appreciated the real gravity that could be. But still, we are passing these lessons on to others who might not know better or don't dare question the Great Master, and these students will pass it along too because it is law. Wow, a tsunami of ripples is generated, it is so! My master electrician, Mark, once told us apprentices that the reason we are still alive (given our history of electrical incidences) is because, for whatever reason—be it luck or watchful angels, that not more than one critical element was present as we absorbed energy from a potential life-threatening event! Yeah, I'd say we lucked out, I know I did. But make no mistake; it's Russian roulette against a noncaring, nonfeeling universal law that is totally indiscriminant to race, creed, looks, money, or occupation. As it is, in our universe, we really are only so much dust, eh, mate!

Mark routinely instructed his electricians that knowledge and experience can breed overconfidence, it shouldn't, but it does. I am sure you or someone you know has said, "No problem, I've done this a hundred times"; a moment later, something unique happens, and then some yelling and whining occur followed by a trip to the Band-Aid box and mercurochrome, or worse, the ER, with its waiting list and overworked nurses! In my case, it seems that every time that I think I've got something in life all figured out, I get humbled in some way. Then I have to lick my wounds and start over.

Nowadays, looking back over my life and activities complemented by my know-it-all competitive nature put me in positions of risk and embarrassment all too often. Openly examining your true motives for things you do or may want to be known for as being pure and saintly can get uncomfortable! Ethics and discipline this day and age are in big trouble as they have been replaced with "if it feels good, do it" and what we don't do today, we can always do tomorrow—that is, if you wake up. Some attitudes and motives are changing, and maybe not for the better, just check reality TV.

As a climber, I am surprised to a degree that I didn't get very badly hurt or killed free-soloing rock and ice, let alone doing live electrical work. I never saw myself as being reckless. I was always very careful and methodical and very adaptable, but so were most of the dead climbers and the burned-up and or dead electricians I've known or knew about. The variables out there simply outnumber our accountability of them all, and this is aside from some threatening boss pushing someone into going too fast or too long or some climber going for fame and fortune! There are the nuts on the freeway these days going 20–30 mph over the speed limit; they dart in and out of lanes creating surprise and panic to other motorists! They may be great drivers, but do they ever think about the other drivers' reaction times, especially when a car was not right next to you, and all of a sudden, one shows up like a missile, and you were about to move into that spot.

What are they thinking? I guess the attitude is that "everything is replaceable"—except life, says the rain on the parade that brought down the blade! The dust that we as humans are gets damned lucky at times; at other times, it is blown into the cosmos to mix with infinity—try replacing that!

One late summer day, I had gone to Little Cottonwood Canyon to do a solo speed ascent of a 5.7 rock climb known as schoolroom west. This climb was part of my solo-circuit routine. I was very familiar with this route and loved it for its diversity. Fueling my fire that day was a conversation that I'd had with a huge bodybuilder named Bill, at the workplace gym! This guy looked just like Mr. Clean (big and bald)! Bill and a few others had watched me climb hand over hand (no feet) sixteen feet up and down a one- to half-inch rope three times with a 40-pound pack on my back! The rope climb was a finishing routine to a grueling workout I did four times a week specific to climbing! After I had finished, another weight trainer named Doug remarked while pointing my way, "Hey, Bill, can you believe that!" Bill then condescendingly said to me, "I don't care, you have no latissimus muscles [I did, and do too], and if you don't let me train you, you will die soloing." Man, now I was sore! I left the gym with that comment stuck in my brain's competitive section! When you are not 100 percent focused, soloing is a big mistake; clearly, I shouldn't have been going to the schoolroom to climb. However, schoolrooms are learning places! While I was lacing up my Boreal Ace climbing shoes, two teenage boys stopped by on their way to another climbing area known as the Green A Gully. We shot the breeze a bit talking about what route they were going to climb.

Then one of the boys asked what I was going to climb.

I pointed behind me and up, saying, "Schoolroom west!"

"Where's your partner?" they asked.

Confidently, I pointed to my feet!

Both kids got real pumped asking, "You gonna solo this thing?"

I nodded, saying I was.

One of the boys excitedly said to the other, "Hey, Wayment, let's go solo our climb." The other kid acted unsure. Before he could respond, I wound up lecturing them about why they should never solo anything, mostly because it is so committing, dangerous, and foolhardy, with very little to no margin for error. In the aftermath, an error could be a lifetime of dealing with unspeakable damage, even death—doesn't that sound fun. Think about this: have you ever heard someone cry for the loss of a loved one who lived maybe into their eighties? It's usually a sad affair! Have you ever heard a dad or a mom cry for the premature loss of one of their children (for whatever reason)? It's catastrophic; an infinite part of them was just torn away! A cousin of mine lost two teenage children a year apart (a son and a daughter)! The pain and destruction and question I saw in the parents' eyes was devastating and without end.

I apologized to these two impressionable young guys for telling them that I was going to solo my route because I was setting a potential bad example! They argued back saying, "But you are doing it!" I told them that it was for personal and private reasons and that I was very prepared for this route even though technically, it is not very difficult, you still could incur a severe fall that could result in death! I then asked the one called Wayment if he had a brother who worked at REI (Recreational Equipment Inc.), which is an outdoor gear shop in Salt Lake City. He said yes, that it was his big brother. I told these two, "I personally know Slater, he is a former Marine and a responsible climber, and I will rat both of you out to him!" Slater would kill them for being that stupid. I made them both promise they would not solo; after all, they had spent lots of money on the ropes and safety gear, so they ought to use it. They agreed then left for their climb. With Bill still rummaging around in the back of my mind,

I chalked up and started up my 400-foot route. I completed the climb in good style, taking about twelve minutes to top out; everything went smoothly, and the experience was very enjoyable. As I was descending the west side of the buttress that this climb ascends, these two kids came by, gave me a cheer, and motored along. Afterward, I thought to myself, *Wow, I still may have sent a bad message despite my lecture!* I also thought about other parts of my life where I took calculated risks to get something accomplished (electrically speaking), where expediency would make all the difference. How important was it really, what was the risk justification even though I was willing to go in the hole so to speak? Hmm…

I had spent most of my climbing career emulating my friend the great free soloist John Bacher and believed I had the square peg in the square hole, and still think so! After all, I was communing with nature in a way that to me was perfectly seamless and natural. I was being one with the mountain, finding my strength, my weakness, and my peace. However, I'd realized I was throwing a pebble into a pond and generating ripples good and bad. I know, especially now, that I scared my family to death, including some of my friends. But my climbing mates and I understood the risks involved, and we were as safe as we could be being who we were—climbers (more rationalizing).

Climbing is not haphazard, yes, it is dangerous, but so is driving a car and so is electrical work or high steelwork! But you can approach these activities with reason and knowledge and skill and have a great experience. Nevertheless, it simply worried my mom sick that I climbed. I tried to explain that there are safety systems employed (I never told her I soloed) that reduce the risk to nil, that driving a car in the city was far more dangerous usually, but she wouldn't hear it. I do remember how her eyes would bulge when she would stare me down asking if I was still doing that dangerous climbing crap. I'm glad my wife never told my mom about my first climbing partner's suggestion: Don had told her to increase my life insurance.

Realistically, a human really cannot control everything that life will throw their way; there simply are some things that we as *Homo sapiens* are too limited to absorb and survive much of the time. Think about avalanches, tsunamis, and major earthquakes that drop tons of concrete on you, hurricanes that hurl you and your car hundreds of yards, flash floods, mines that cave in, and lightning strikes—nature et. al. We have also created some dangers in our world (e.g., trains, planes, and big rig diesels; just have any one of these run over or into you and see how you come out of it!). What happens when you fall climbing or slip off a ladder or out of a tree? Did you know that you fall about twenty feet in two seconds, and at the twenty-foot point, you are recording thirty-eight miles an hour? No kidding! I apologize, but I must lay an old axiom on you, "It's not the fall that kills ya, it is the sudden stop." Well, we can certainly make prudent decisions that can help level the playing field. But alas, we are humans—complete with a head full of psychological dilemmas that can and does sway our decision-making process. We diagnose what is important to keep us fitting in, then in flows the temptations in life—fame, sex, drugs, money, and then material wealth, along with the ability to wield power over others, etc.! I know it all sounds crazy, but am I wrong? I wonder all the time about the balance between the good and the bad. Sure, free agency is each individual's right, but at some point, responsibility for your influence or action has to be weighed in and accounted for too. I'm not a fan of the constant use of "definites" simply because of the amount of variables we face and find that I usually wind up becoming more flexible in my judgment of any one scenario. However, some things are always right or always wrong; *always*, they have withstood the test of time and tribulation.

Not too long ago, a friend fell while soloing, dying as a result! He had chosen to continue doing what he loved even with a significant injury that compromised his capability and safety. Some folks say he was stupid, others say he was doing what he was designed to do. Everyone had an opinion! He left a family and many friends and admirers to live with the memory of him. Sure, he could have died in a car or plane wreck; accidents happen all the time, but when there are known odds working against us and we go in against the risk anyway, what was the priority? Now, I'm not talking about war, where you at times have the odds stacked against you (yes, even US soldiers). In this scenario, you and your buddies have to survive against an armed enemy who wants to kill you. Here, you will be forced to work and think in and out of the box, to adapt or die.

Note: I am not trying to be a judging entity here, I am merely (albeit directly) giving you something to think about. The question becomes, what the hell is driving our decision-making process? What are our priorities? Are our personal goals and desires and motives clouding our ability to reason? Isn't this like being on drugs? Can someone in this state take the time to consider what they are about to do, cause, and/or change? The bottom line is this: when the train slams you and your car, you (no matter who you think you are) are like the ant that errantly is under the boot that just squashed it.

One night at the Air National Guard Base in Salt Lake City, the linemen from our civil engineer units electrical shop were about to reenergize a twelve thousand–volt switch they had been working on. Unknown to these technicians, a bolted fault had been created (which is a solidly connected direct short) within the switch gear during repairs. As this main power switch was closed (turned on), a massive explosion occurred, blowing the two unlatched three-by-six-foot doors wide open, sending a huge arc flash and blast out of the housing. Amazingly, no one was in the blast's pathway; thus, nobody got physically injured! I have often wondered about the mental impact that may have been induced into someone though? Remember Ed running from the mechanical room when the air compressor turned on, screaming its brains out? Ed was scared to death!

Being involved in or around intense events like this has their unique impacts on the individual. Some people seemingly have no reaction nor are they impacted whatsoever, others melt like ice cream openly, yet others hide it away to be its slave in their own darkness. "The red gods of fate were on those linemen's side that warm summer night long, long ago!"

This day and age, the NFPA (National Fire Protection Association), OSHA (Occupational Safety and Health Administration), MSHA (Mine Safety and Health Administration), the NEC (National Electric Code), and risk management have finally recognized that there is an alarming number of accidents happening daily in the workplace across these United States. Our burn units/ICUs, and morgues have become residents for our qualified and unqualified electricians, HVAC techs, and other tradesmen and workers who work in unsafe environments and/or execute unsafe work practices. Accidents come about because of a myriad of reasons; corners get cut, and we (the worker) take our so-called calculated risks. Sometimes we win, and sometimes we lose. Think about the unqualified workers, which is anybody that has not had specific book learning to include on-the-job (OJT) training relative to the subject and is considered incompetent to the required level to perform work on said equipment! These folks sometimes get pushed into situations that simply are an outright crime, and sometimes they volunteer! They are lucky, lucky, lucky if they overcome the obstacle without ruining the equipment, hurting or killing themselves and/or others too. They will eventually, however, come face-to-face with the laws of physics, and life will take its course. Again, all this occurs because somebody wants to save time and money or might be trying to impress someone else so as to keep their own job (a.k.a. saving one's neck). The maintenance field within some government agencies is particularly at fault in this regard! It is in need of a serious evaluation in terms of how they hire, utilize, evaluate, and train their personnel! Amazingly, government and civilian agencies who may create very appropriate safety guidelines and SOPs (standard operating procedures) and employ thousands of tradesmen, general maintenance workers, mechanics, drivers, engineers, managers, housekeepers, etc., at times seem clueless as to what is happening on the ground in regard to these processes within their organizations. The leaders talk the talk and may even be able to show trends and forms and scribble! So how do they manage to survive (even for years) yet leave no footprints? Here are a few examples!

A couple of years ago, my partner Joel and I had just returned to our office at the National Guard Headquarters in Draper City after working for several days at an armory in southern Utah. While I was walking up our warehouse hallway, headed for our gym to stretch out, our carpenter came out of one of the warehouse doors. He looked kinda stressed and said he was glad to see me, then asked if I could help him with a fluorescent light fixture. It was quitting time, but he seemed to be anxious, so I said, "Sure thing," but was curious as to why he was dealing with a light fixture. As we made our way into the work area, the carpenter explained that our foreman (the "two-minute safety class" specialist) had assigned him to work on these warehouse lights. Understand that these lights, which are suspended sixteen feet off the concrete floor by simple jack chain, are very unstable and are powered at 277 volts and are very tricky to work on! The carpenter said that he couldn't get the light to work even though he had managed to change one of its ballasts and didn't understand how to wire the other ballast, which looked very different from the first one. The fixture covers, which are a drip pan / reflector style typical of warehouse/garage lights, were still off; so from the ground, I could clearly see all the components. As it turns out, the ballast our man couldn't figure out was not a light ballast at all. It was a battery backup, which is designed to light up the bulbs when the power goes out. These emergency backups cost about a hundred bucks, and if connected improperly will be burned out! I asked the carpenter if he had turned the power off before he went to work on the light fixture. He pointed to a light switch on the wall and said that the foreman had told him to simply shut off that switch and he would be okay. Wow, what an assumption! I then asked the worker if he had checked with a volt meter up at the light to make sure that the power was off; he shook his head and said no. I pointed out the fact that the light switch he was told to use had nothing to do with the light fixture he had been working on; he responded with a look that was one of surprise and embarrassment. I also pointed out that his ladder was at least one size too small for the job (he used the second-to-the-last step to reach the light) and that a small platform lift is normally used for stability to repair these lights. Along with these issues, he had worked on the fixture with the power on and thought it was off and was damned lucky that he didn't get zapped by 277 volts. This man would have fallen about ten feet if a shock caused him to lose his balance while wrestling a fixture that was moving around on chain supports! I let him know that he could have been killed (people die routinely falling off ladders from this height) and would have for sure been badly hurt. I informed him that he has the right to say no in a situation like this. This person told me that he doesn't like working on electrical and that it scares him anyway. It's worth mentioning that our foreman had absolutely no electrical skills whatsoever and that our company got real lucky on this close call, not to mention the worker avoiding injury. I am amazed at the outcome because of all the different negative elements present and that we all walked away unscathed, it was a case of pure and simple blind luck! I'll tell you what though: in life, you gotta be careful 'cause good old Murphy, he's watchin', and he'll getcha because time is on his side.

Back in the early nineties, while working as the base electrician at the Air Guard in Salt Lake City, I got a call from our work control office. Sgt. James, our work control manager, asked me to go to our motor pool to assist an electrical contractor find a 120-volt power source for an HVAC unit. I thought, *Wow, this (I assumed) journeyman or fourth-year apprentice can't locate power on his own, especially in a nearly brand-new building?* When I walked into the electrical/mechanical room our man was busy working in, I was slightly stunned by the drama before me. The electrician was a kid about fifteen or sixteen years old; he was hustling around like a nervous cat boring holes through the sides of electrical panels that were mounted side by side. Of course, most of the panels were live. Some were rated at 208/120 volts, and some were 480/277 volts at 60 hertz, and there were several 400-hertz panels that were, at the moment, not energized, so I asked him what he was doing. This kid's goal was to establish a circuit to a window-mounted air conditioner some 60 feet away in an office. As it turns out, he had entered the first 400-cycle panel with his conduit from the HVAC unit and realized that this panel had no power because, in his own words, "It appeared to be hooked up to the little generator in the corner (a 400-cycle converter) which wasn't running." He thought that it must be turned off, and he couldn't figure out how to turn it back on. So to make things easier, he bored a hole through the side of this panel into the next panel, which he discovered was also part of the generator that was turned off. He continued to bore a hole into the next panel, which was a live 480/277-volt 3-phase 200-amp panel board. He told me that this panel was wrong too because a mechanic from the shop had said so; the mechanic was right too! So he had drilled into the next panel, which had the power he apparently needed (it did). As he progressed from panel to panel, he installed pieces of EMT (electrical metallic tubing) conduit (crooked) through the holes he had bored with each end of the tubing protruding far enough into the enclosures to put an EMT box connector on backward securing the conduit. I guess his heart was in the right place at least!

In each panel, except the 400-cycle panels that were deenergized for the moment, the kid had came dangerously close to live parts and didn't know those parts could harm him! I explained to him that he had all the protective covers off every panel exposing all the live parts. It was pointed out that a simple slip one way or the other could have been catastrophic and would have been bad for him and the many other people who would have, out of necessity, become involved in a big legal battle over the death or injury and, to a lesser degree, the property damage. I had to stop his work from proceeding, and as we talked, I asked if he was an electrician, he said he was, so I asked him for his electrical license, he said he didn't have one. I told him that by law, he needed to have a license to practice. I also told him that the generator was a 400-cycle converter; he responded with, "A what?" I then asked who he worked for, so he gave me the HVAC company's name and then told me that he did odd jobs for these guys when he wasn't attending high school, where he was about to be a junior. Kill 'em young, I say, and for money! So I called his boss and told him that I was shutting this job down until a *qualified* electrician was sent to complete the task, not a five-dollar-an-hour helper. I also let him know about all the risks that had been taken and damage done to our property and would put him in contact with our colonel and base engineer so they could iron out the offense. This was clearly a case of a company cutting corners for profit's sake, meanwhile violating regulations from the State of Utah, OSHA, the NFPA, and our contract with the HVAC company. In the electrical trade, as in life, many things are still learned the hard way! Even today, as we move toward learning to use proper PPE, or personal protection equipment, and attend regular safety classes and seminars, the general mindset is one of accepting unacceptable risks! Realize that the worker is just as much to blame as the employer. Why? Because you as an individual are just as responsible for your personal safety as your employer; after all, it is your well-being and ultimately—your life. We do this without ever taking into account all the associated fallout from an accident, or if we do understand it, we are that lame to think it won't happen to us or at the places we work in. Someone attacked the Twin Towers! Whoever thought that someone would attack the badasses in the United States, and right on our own soil, really! We are not immune to anything.

Look at it this way, someone, OSHA for example, is trying to force us to stay out of burn units, ICUs, traction, casts, and graveyards, what's up with that! I just don't understand. Why would a boss, for example, want to pay the expenses and fines for an injured or dead worker? What about having to face the law over it or his own bosses, who will hang him or her out to dry? Oh yes, they will! What about facing the injured or dead person's family? Eventually you will even have to face yourself too and the long days and nights of your mistake and all the people that it affected. The ripples will go on and on and on, and you may rot away in your own self-pity and ashamedness and even die more miserable than a rat long before you ever get an ounce of solace. As a matter of course, however, a boss rarely, if ever, has to work in the environment and under the conditions of his/her subordinates! So to a desk and computer impersonal boss, an employee working on a live electrical circuit or a high-pressure steam boiler or laying asphalt or being an on-the-ground cop or fireman or an ER nurse, etc., well, things ain't so bad, it's passé even.

In this boss's mindless mind, the situation usually distorts to this idea: "Besides, we've only had a few accidents" or "It's never happened before, so quit overreacting and get to work, criminy, I'm a dollar waiting on a dime!" I know that not all bosses are this cold and/or stupid, but some definitely are, and people are but chess pieces to them—whoops, did I just make some of us sound like "subcreatures"? Hmm!

For the final dose on this topic, just look at military history! We militia fight and die for your and our rights and freedom! The enlisted man is the iron that makes up the military framework! Yet only until recently, we were but a number and were considered quite expendable and replaceable—uh-huh, just a necessary evil. That's right and damn straight! These days, we are respected quite a bit better by the professional side of the house (officers), and it was earned the hard way.

A few years ago, I was helping a general (who was in charge of Homeland Security) with understanding espionage processes relative to electrical systems.

The general said to me before we got started, "You were sent to me because you are a direct and honest on-the-ground master sergeant."

I replied, "That is correct, General. I also train, work, eat, and sleep with my troops." I went on to say, "I like working with the facts as much as possible as it speeds up my output of useful productivity."

"Good," replied the general, then he said, "now I want to know what you think of generals."

I replied, "You don't wanna know!"

He fired back with, "By god, I am a general, and I want to know!"

I told this general officer that I would tell him, but it shouldn't impact my career; his eyebrows reached an exceeding height, but he agreed that I would be spared! So I asked him if the butt print he left in his office chair was deeper than one and a half inches, he asked, "Why?" I then told him that if it was, that he was in it too much! He asked what I meant. So I told him that he did not have a real clue as to what went on in his organization, especially at the bottom with the lowest-ranking people! Gruffly he said, "What?" Then fuming a little, he said, "I think you're insolent." I told him, "You are right," then proceeded to say that it was his staffers that kept him up to date on the conditions of his house (which is appropriate most the time) but that they may not always tell him the facts. I reminded him that I had been in the service for about thirty years and had been deployed all over the world and understood very well exactly how we operate. I have not been keenly happy with our leaders' assessments of our condition and conditions from the armchair or by some floozy trying to cover his/her butt. I suggested that in between cribbage games that the general ought to go hobnob a bit with those people at the bottom, that it would do everyone a lot of good and maybe promote the feeling of being part of the team etc. The general said he understood and promised to follow through. I hope he did!

I seem to be going on and on about this safety and survival thing; it's because life is a blessing, a precious thing. If you think I am wrong, just look into a newborn baby's eyes, can you put a price on that? There are good protocols to follow that may prevent accidents or at least reduce the impact; yes, it can be expensive and inconvenient, but what does a life cost? How much does an injury cost—come on, in dollars and centavos?—to include the emotional impact that happens to coworkers, friends, and of course the family! What is the total cost of these losses?

Long ago, while surviving in a burn unit, I learned a few somethings. I found that it was easy to become lost and that there were levels of pain that can't be accurately described by that word (just ask any third-degree burn victim how they feel about what has/is happening to them or a suicide victim up to bat), and finally, there is always at least one person who needs to see you make it. Having faith in the Creator can really help you cope with the infinite challenges you will face in life and death, I promise.

9

Common Sense in the Fast Lane

This is another element relative to our safety and survival topic that needs some attention! I'll say it again, *bon sens*, which is French for "common sense"! Some people are born with it, and some will never realize it; some folks may disagree, but it can be developed to a very high degree requiring some hard work and discipline and awareness.

Have you noticed that plain old good judgment is going out the window as our world becomes more and more diluted? One popular catchphrase is, "If it feels good, do it." Everything seems to be "okay" if it is free and/or easy. We simply can't take this approach with every single situation we come across day to day, that is, if we expect to do very well. Obviously, the topic of common sense and how it applies can become quite involved since it covers a wide range of elements. Of course, it is very difficult to make perfect decisions every single time, so to shorten things up, just try to be reasonable and as practical as possible.

These days, we seem to rely on electronics for all the answers. I personally enjoy what computers do for us, but it is not the end-all answer to every question you and I will have. Of course, technological advances have made our lives so much better in every arena you can think of. We have more ability to improve and save lives than we ever had. Cars for example are so much more efficient, jets get us to and fro in minutes, and we humans will visit Mars in this century, and the humble computer is smack-dab in the middle of it all, speeding up all sorts of processes. Just the other day, my dentist added some volume to an implanted tooth to stop food from getting stuck where a gap was. I noticed all the unusual-looking tools he used, especially the device to keep my mouth open and block debris from going down my airway; I was quite comfortable for the thirty minutes he had my yap open. A computer helped develop these precise and efficient devices. However, we create these great things at such an alarming rate that before some new technological device is getting cold on a store shelf, it has been replaced with something better, or should I say newer. Unfortunately, we are in such a hurry to move forward that we don't realize or see the possible negative impact of our creation, we just want to go faster and faster and get better and better, to hell with the fallout.

Newton said that for every action, there is an equal and opposite reaction! Since we don't always watch our back trail like we should, and don't notice what is off to the side or what has just passed us, as a result, we get lost and into trouble, am I wrong? Think about road rage. The nature of electronics has helped create a people that are entirely in too big of a hurry—be it for their personal or professional lives. We conduct all manners of business behind the wheel, claiming we can multitask just like fighter pilots; I am surprised there aren't more accidents and fatalities. Think I'm wrong? Cell phones (the privacy destroyer) are helping to fill up ERs and keeping auto-body shop owners busy. Law or no law, we are creating an unfortunate variation of job security. Like nature, electronics can be a friend or an enemy; we just need to understand its limitations. People also have limitations, so I try to trust my intrinsic senses and instincts; both were derived partially from lots of schooling and OJT. I have found that my gut instincts are usually right if I will just listen. I also think that computers have driven us indoors and secluded us from the natural world and maybe desensitized us to a degree from our basic senses and feelings. Could we be becoming less human as a result? The bosses I have been picking on might be a little victimized themselves by technology. As life evolves and speeds up, could it be causing some of their rash behavior? Again, tell me that we aren't in a huge hurry and more than we've ever been; we're so busy we barely even know our next-door neighbor. I still say, however, that we generically understand right from wrong, and most people have some modicum of common sense, that is, if they will stop to think for a second and be reasonable. I also think that treating others with respect and fairness is still and should always be status quo.

This great tool the computer is awesome, and of course it helps simplify things that smooth out life for us, but it has simplified identity theft too. I really appreciate the hackers for all the stress they cause as you can't even use your credit card at a gas pump for fear of being hacked. I want to call these hackers some name associated with lots of profanity, but no one word—or groups of—covers the way I feel. I guess so long, as the newest, biggest, baddest brain has power (that is, electricity), we all win, and the cyborgs are happy. But have we become a little too reliant on these portable "think for you" tools? How did our grandparents survive without this technology? Think about all the people born preelectronic age, and all the pros and cons they have witnessed! Do you have a backup plan for cyber failure? Better be prepared because it does and will happen, again and again. I also wonder if the cyber world has not helped create a reality crisis. It seems that after the computer landed on earth, people began staying indoors more, getting all wrapped up in the fantasy computer world. Then the idea of "if it feels good, do it" came about, thus ushering in a very carefree attitude toward things that help keep a society together, like sound ethics. Many things this day and age are too easily explained away and thus fixed with a pill and/or a keystroke. I find it less than humorous when servers go down and put the world on hold. But it's downright scary too because this tool is not going away, and as progress will have it, we will become more and more reliant on its brain rather than our own. Because of our dependence on electronics, it could become easy to get backed into a corner that is potentially inescapable! Better have a backup plan of some kind (i.e., food storage, extra cash that's not in a bank, hand tools, survival skills, etc.). Remember the sheep story? When one goes over the cliff, the rest follow; better have some common sense. Still, people just have their mugs constantly stuffed into their smartphones and typically won't say hello or look at you when you pass them on the street. (I almost hate my phone.) Are we that bored or what? Am I too antisocial, judgmental, or what? I know what you're thinkin'—pick, pick, pick.

I suppose that's because I'm a hands-on guy and come from a more mechanical upbringing. I may appear to be somewhat closed-minded to technology, but I recognize its value and use it as appropriate. I am also very impressed with the incredibly creative minds who are driving it forward! For instance, in the medical world especially, we have MRIs (magnetic resonance imaging), which can show just about every level of the body one can imagine and can help doctors locate and repair problems they couldn't see at all before. As an electrician, I can understand the struggle of not being able to see inside of a body accurately enough to determine the cause of a problem; I have faced similar dilemmas with electrical systems. But with such advanced equipment these days, can you imagine what the future holds? Maybe one day, they will be able to get into the all-illusive mind, cure cancer, and create safer implants, prosthetics, etc. I just hope these great minds who create these tools, including our medicines, will pay more attention to the bad created by the good, like common side effects that have become so commonplace. Over the last decade or so, we have identified many new diseases, some of which are questionable and appear to be misdiagnosed and abused—at least on occasion (e.g., ADHD [attention deficit/hyperactivity disorder]).

Many kids and adults are diagnosed with this problem, and at times, it seems that the real problem is they need to become better understood and taught to be more disciplined! I question the diagnosis at times because some people never skip a beat with the iPad, smartphone, or Wii. It seems that if some of these ADHD victims like something, they stay right on track as opposed to finishing dinner, yard work, or other things that are not their idea of fun! I personally taught one ADHD kid how to do a backhand shot in the game of ping-pong, and after twenty minutes and a few refreshers, he was an expert at the maneuver, pretty amazing that he could focus long enough to become so competent so quickly! Some people or organizations attempt to use these diseases as reasons to fail in academic and work-related activities and are blown away when the excuse/reason is not accepted! Eventually I would hope that people realize that whether or not they have a real problem, they can achieve their goals!

Some doctors are not helping American children or adults in the way they should be and may have added to the problem with some of these lackadaisical assessments, which have, by the way, penetrated many other levels of medical issues. Simply put, they are not thoroughly troubleshooting their patients, which I call laziness! Again, they seem to be in a hurry, and the approach is to get you on medication or therapy (granted, sometimes, it is the correct approach), or they might even have their physician assistant (PA) treat you, doing their light work—which is understandable at times! However, a PA was getting a little too involved making decisions that my surgeon should have been making! So in my theatrical thinking, I often morphed to this idea: Oh great, a PA is gonna assess me! So I thought, *Okay, I'll get you back, I'll just have my accountant wire your house, ya dammed yahoo—and for double my normal fee.* Terrible thinking, creating vicious cycles—stop it! Maybe though, doctors are facing so many new problems that they readily do not have good answers for, and maybe the sheer number of these issues outweighs their ability to address them all. Waiting rooms are as full as they have ever been, and most physicians do their very best to help their patients, I believe that! Think about what they endure despite their efforts! Nonetheless, it's obvious that there is a serious ethics shift in our country that may be driven by the "easy and faster, I want it now" idea! I personally have been seeing way too much inconsistency with many of the diseases, implant surgeries, etc., so I wonder about reality. Hey, I'm in it with ya, but there is a reasonable limit to everything, there is even talk about when a person should die versus the money required to keep them alive! Some people think they belong and others shouldn't be allowed oxygen. We chase oil and money, and some of us live big and some get by and some get swept away, and sometimes it's because the "some" were cyber illiterate. Man, Jesus has his hands full.

In my military unit, the older military folks began noticing a distinct relaxing in the recruits and the regimens of boot camp around 1980! The concern was that a dangerous attitude was forming, and eventually we would end up with soldiers who were weak in discipline and may not follow orders. Some people wouldn't join unless the government let them have more lamb! So these days, the military waves big sign-up bonuses in front of everyone's faces, knowing some people are willing to die for money. Whatever happened to duty to your country first? JFK (love him or hate him) would flip out if he saw this crap taking place! It seems that some folks want to get to paradise without making the journey. Living in the United States has allowed me and everyone else to have more freedoms than anybody else on the planet, and we take serious advantage of and abuse this privilege of ours, but freedom is not free or easy! We have so many choices that it's hard to choose without going into shock! Criminy, my watch is a wonder of design and function and is awesome once I get through its bookful of instructions—there I go pickin' again—I actually really like my watch! However, I'm glad to be a basic human with feelings and a mind free to be creative, happy, and mad and independent. I'm lucky to know how to adapt and to have been born and to live in this great country of ours! It's good to know that I am an asset to an organization (or have been, now that I am retired) and have been willing to please no matter how many times I am beaten and told that I don't do enough and am rated a zero in performance reviews. Yep, I came from the brutal authoritarian era, so I should be used to rough treatment, but nobody is perfect. Actually, being a zero does get old for me, so at least if I can be a plus one or a negative one, this would mean I am moving somewhere and avoiding boredom and stagnation! I guess I should explain!

I worked for a *pair* of supervisors who ruled over people in such a way as to always try to make them feel inferior; the threat of losing one's job was ever present, and the stress was terrible! You simply could not please these idiots. In one year's time, they caused—one way or another—twelve of fifteen employees to leave the civilian workforce of our military office and replaced them with what I will call friends. In the end, two other coworkers and I survived these caustic-minded freaks (I will go into more detail later). Seriously, the way things were, it seems that some of our government offices would make great POW (prisoner of war) camps for Iraqi or Taliban soldiers; soon they would realize just how good they had it at home.

It has been said before that change is a constant in life, you can count on it! So learn to be patient, mindful, and have faith in the fact that goodness will prevail! The mean individuals I mentioned were removed from our office and reassigned to lower-level positions. See, what goes around comes around sometimes!

Despite this drama, of course I have been well taken care of too! With the bad, there is always the good; it's one of the laws of the universe: yin and yang balance in the atomic structure, yes sirree bub. Know this: we were created by a greatness that happens to be an object of intense debate and scrutiny! No matter what though, we are destined to evolve, to multiply, succeed, and to love or destroy and kill if we choose. We also possess minds with the ability to reason and to divide and conquer, hopefully in a productive and fair way. In the workplace, we are usually there longer during our waking hours than we are at home, so why do we tend to have so many issues in terms of human relations? I am sure that psychology departments wonder if they will ever get to the bottom of that topic. We could help by being more flexible with one another and focusing on our responsibilities to the best of our abilities, and that means everyone from the top to the bottom, even the president! However, once again, the human element is so diverse with all of its associated paradigms that...could it be best to live and let live? My nature is one of intense organization and output! I know that my coworkers and mountain-climbing partners have benefitted from that quality, but have suffered from it as well; my personality or nature asks a lot and demands even more. I am murder on myself in everything I do, relaxing to me is a swearword; however, if one is not careful and attentive and understanding, one tends to levy these same demands on everyone around them, which can be very unfair and destructive. I learned part of this lesson from younger climbing partners like Keith, who was a physical animal who knew no bounds when there was a goal to reach. I think Keith could have won the Iraqi war all by himself; he is that intense and capable. This man exudes the idea of mind over matter and owns the physical brawn and skills to back it all up! My "normal" backpacking and climbing partners, being humble folks, only quietly mentioned the pace of our trips together. I wish I could have heard and understood their thoughts so they might have had a bit more fun, so we could all reach our collective and personal goals. This energy washed overboard into my workplace too, I always had to win! I had even given myself permission to die while trying. You may not like hearing this, but everything in life has its limitations, and understanding this law is a key to everyone's survival.

When I started climbing and hiking with Keith, I realized I was aging; that is, compared to him (an Ivan Drago twin), I finally understood what I had put my other buddies through who were normal guys, not driven athletes or tradesmen who compete with the whole world. In every walk of life, there are some simple rules to live by, and we should already know a few! Think about this next statement for a minute: Your team is only as strong as your weakest member. I had thought of and repeated that saying often (I liked it) but never paid attention to its trueness as much as I should have, that is, until I wished for Keith to employ it for my sake! Keith was more forgiving than me, however, but our credo was, "the more you sweat in peace, the less you bleed in war," and that was that—"no pain" was the rule we simply trained to survive disaster. For some people, this practice isn't too fun to live with, but in the right usage, it may be necessary (i.e., war, Antarctic exploration, etc.), but in the wrong situation, it's not a good idea. I do believe, however, that there has to be an average that is achievable by folks in any walk of life. Of course in any venue, there will be overachievers and underachievers by comparison.

The Boy Scouts have a motto—Be prepared—and the Boy Scout oath! Both of these guides contain simple processes that will help get you through life. These guidelines plus the Ten Commandments seem like a good benchmark to use on how you do business with yourself and others. I believe these guidelines can help someone manage just about anything successfully. If you watch someone who has mastered these practices, notice how they work with others, also pay attention to how they approach hurdles (everyone has them); they often find a reasonable solution that works for everyone, and then they move on. The idea is to help yourself and others to have a "smooth as possible" life no matter what the circumstances! When the military is involved in a war, the soldier and his buddies are often at times all they have; it's where love and respect can dictate life and death. Duty and honor are unconditional and state that you may have to go save someone else's life; it doesn't say *you* are coming back! Accomplish this level of caring, and you ought to be very happy and at peace; it doesn't matter if you are the boss or the worker or an aunt or an uncle. To get love and respect, you have to give love and respect with no thought of anything coming your way in return.

Someday you may even have to endure a real-world disaster like Hurricane Katrina. What if Y2K had happened? I believe it would have been an event of epic proportions: "Holy cow, no electricity in the middle of winter!" An event such as this is hard to imagine, let alone surviving it, and maybe for weeks or months. Just think about how many people would perish—and right in front of you. I can tell you that as the Y2K potential epic was approaching, I had dozens upon dozens of worried people come to me to install generators at their homes! People were telling me about the big full-load generator they had already installed themselves, which means that the generator would handle the electrical needs for their entire home! I asked this second group if they had done a fuel calculation because a generator that supplies that much power will be fuel hungry. I asked how big the fuel tank was and how they propose to get more fuel with the power shut off citywide etc.? Generators also require routine maintenance, so what about spare parts and the knowledge to perform the work, all in the dead of winter! I also asked both groups how much ammunition they had.

"Ammunition? What for?" someone worried.

"Well," I said, "some cold starving people will want what you have, and you will have to defend your valuables at some point, maybe the walking dead will come by—I don't know—just be prepared." The question is, can you be in control long enough after the world tips over to sort things out, get shelter, and build a fire? Can you? Can you do what someone so desperately needs you to do when there is a bad car wreck right in front of you? Or will you complicate things and let panic erode your limited reason until you become totally at the mercy of the merciless situation right when someone needed your help? It's all on you, alone against the laws of the universe, and your boss can't help ya, He's stuffed behind his computer, and it quit working—game over!

Can we be just a little more humane to one another and realize most everyone has value. I say *most everyone* because our great torturers are lost entities and are the great Creator's problem.

I've been pretty peeved about lots of things for a while, and maybe I expect too much and judge too much—it's so easy to point fingers instead of being productive, right? Maybe we all need to step back and get back to basics!

Lastly it's the fall of the year 2013, and our government, who can no longer look their trusting taxpayers in the eye, have managed to not be able to make a decision (again) on the hot topic of ObamaCare. So when time ran out, with the deciding parties divided, we incurred a federal government shutdown. This action affected the lives of countless federal employees, except the rule makers who wouldn't make a decision concerning our citizens' future and well-being. What about the additional stress on the troops overseas? For the rule makers, it was just another meeting and day at the office. They have no idea what life at the bottom is like; they will go home, complain to their spouse, and maybe work a few Saturdays just to have the other party disagree again! So they take an aspirin and partake of the dinner the house staff has prepared, and it's off to bed coddled with silk sheets that you and I bought! Amazingly, 2014 threatened another similar federal shutdown episode…what is their problem?

In lieu of all my ranting and crazy directions of examples—bare minimum—remember this next bit!

The decisions you make and the actions you take—to include the words that fall off your tongue—no matter who you are, affect somebody else most of the time, and these effects can last a very long time and do lots of harm. Our country is a great country, but I feel that we are slipping into mediocrity. The leaders we vote into office need to remember what their assignment is, which is to do their job and to remember that people that make ten thousand dollars a year are just as important as someone who is in six-figure income brackets; we all matter! If the middle- and lower-class factions don't matter, well, that's a great way to divide the country! There is no place in these important offices for personal agenda and/or worries about personal gain. Remember, "united we stand, divided we fall."

I hope to give this message to our government leaders in office:

Also remember, we voted you into your position to not be the cream of the crap or the top of the bottom or the best of the worst, but to be caretakers of our (not yours) great nation and people! I wonder, when you stand in front of the mirror, do you realize how much influence you have over so many lives? We as a country need you to remember who we are, the citizens of the United States, and why all our forefathers came to this land. We need you to make timely well-thought-out decisions (it's not that hard) especially when it's going to directly and quickly affect so many, *or* leave your post. We pay you far too much money to be so derelict; we'll find someone who can do your job and will be more reliable.

If I have caused anyone, whether you are a government leader or a grocery clerk, to wonder if there aren't a few things they need to work on in their life. Well, good! No time like the present to get your issue(s) on the table and decide on a corrective pathway. It may not be too comfortable though, I am sure there are folks counting on you to stick with it till you get it fixed. You and probably those dependents will feel so much better, I promise.

Know this: no matter what your challenge is, the clock never stops; and if you wait five years, your issues will no doubt be bigger and harder to resolve, but it's all up to you and the free will that somebody divine gave you.

I worry though, where are we as a people and a world headed?

Storms

10

The Warning

My life, maybe by design, has been surrounded with challenges that were very physical in nature, including waking up in hospitals too often. There were also the psychological dilemmas too, like learning how to learn and learning how to deal with people and all the different personalities that are in our world and finding reasonable solutions to make things work. I wondered at times what God was preparing me for or, from a different perspective, had against me? To be fair, I am aware that I have tested his eternal patience to the limit. I also have realized that many of my challenges and the way I approach life have been intertwined and thus complicated by my upbringing and attitude toward the earlier part of my life.

Life has a funny way of doing about-faces, which is a military term for some cool footwork that makes your backside appear where your front side just was or vice versa (as the case may be) and could be just what the doctor (mostly proctologists) ordered.

It was 1989. I had four children ranging from fourteen down to four years old and a thirty-three-year-old wife depending on me. I had just made a big change in my workplace, which included a commitment to further my education in the electrical field as well as assuming a lot of responsibility. I was also right in the middle of my military career too! My plate was full, but life wasn't done changing things up and was about to deal me and my family a heavy blow.

At this point in my life, despite my other commitments, I was spending a lot of time in the mountains camping with my young family and was climbing rock and ice as much as possible. During the month of July that year, I wrote this poem:

Life passed the mountain in footstep
Few words were spoken
Vistas were etched into the soul
The climb was there it could be done
But a card had been withheld
The Joker, Alpine weather
Today the ever-silent mountain would win
A hand dealt with unemotion an irreconcilable force
Ropes winged to rappel
All for a split second was explosive
A shaft of death found its fate
The silent mountain again stands in wait
Timeless…

My wife had once told me that she believed that I would be killed in the mountains, and I am sure that my first climbing mentors fueled that belief by telling her to increase my life insurance. Teri only saw me climb once, and of course, I was soloing. I don't think she understood how dangerous climbing could be; in the end, she simply asked me to "Please, always come home!" One year later, nearly twenty-five years ago now, I was on a backpacking trip with my electrical shop supervisor SMSgt. Mark Hendrickson and our two sons—both named Eric; my Eric was nine years old, and Mark's son was ten. We were headed to Blue Lake in Naturalist Basin located in Utah's High Uintas Backcountry, formally known as the Primitive Area. Two weeks before we were scheduled to leave, I had gone to the Salt Lake City International Airport to see my climbing partner off; Jay was headed to boot camp to become part of our guard unit and a firefighter. Before Jay got on the plane, he pulled me off to the side and told me to be careful (prudent partner) out in the mountains and to pay attention to my gut instincts; he had had a feeling that something was gonna happen to me while he was gone. I promised I would pay attention to the inner voice and bid my partner farewell. About a week later, my business partner Rick and I had drove to Reno to meet with a business expert named Don and show him our plans for starting and operating a backpacking business.

While we were talking, out of the blue, Don asks me, "Layne, do you know what one of the most important possessions in life is?"

I hesitated.

Helping me, he said, "Time."

I replied, "How come?"

He finished with, "Because you don't know how much you have left."

I didn't realize how ironic this statement would become!

Mark, our two boys, and I arrived at Blue Lake in Naturalist Basin on the afternoon of August 7, 1990, set up camp, and settled in. The hike to this beautiful basin is rocky and pretty steep in places, but it is a bit of paradise, and it was great to be here with friends. We fished, ate some food, and took in the awesome views that were in every direction, including up. The next day, August 8, while everyone else fished and hung around the lake, I hiked up to the east ridge of Mount Agassiz (12,240 feet) to have a look into Middle Basin, just north of our basin, thinking we may want to visit it sometime! Remote and beautiful, Middle Basin is surrounded by four huge peaks that encircle most of its area, displaying an obvious glacial curvature carved eons ago. This cirque also contains some popular fishing lakes, Ryder and Mcpheters, plus, it really gives you the sense of being far away from it all, which 99 percent of the time is a good thing. In the meantime, I took a self-portrait sitting by a rock cairn that marks the high point of the ridge, with massive Hayden Peak in the background!

I had dreamed of climbing even bigger mountains like Mount Everest and K2 in the Himalayas and had been preparing myself for such challenges if the opportunity should come about. Later that day, Mark asked me how the view had been.

I replied, "Incredible."

He then asked, "Will you take me up there tomorrow?"

I nodded and said, "Sure!"

So we had dinner and watched the sun set in this little piece of heaven then slipped off to bed! The next morning (August 9, 1990) when I woke up, I had a feeling that something wasn't right; my gut was telling me to leave this area. A few minutes later after my son Eric had woken, I mentioned my feelings to him; he said he was good with leaving. A little bit later when everyone else was up, I went over to Mark, who was now sitting by the campfire ring, and told him about my instincts. Mark wondered if I hadn't just had a bad night of sleep; after all, I was still using half-inch-thick sleeping pads, which are not known for comfort, plus, we were on our vacation, and there wasn't anything that remotely seemed threatening. I reluctantly agreed, but Mark did say that we would keep an extra eye on things, and if anything comes up, we would pack up and go home, stopping at Dick's Drive-In for a burger on the way of course! Nonetheless, the nagging feeling would be ever present the rest of the day. So after deciding to stick it out, we also decided to go up on the east ridge of Mount Agassiz to have lunch and take in the view I had experienced the day before! Interestingly, I had gone the day before based on an instinct; the message was that if I didn't go that day, I might not get to, at least on this trip! I can say I have missed several opportunities in the mountains because I thought that I would get the same opportunity later, like seeing a moose or some other interesting creature—they don't hang around waiting to get their picture taken.

While we were all up on the ridge enjoying our time, a couple of men came down from the summit of Agassiz; their names were Jack Booth and Larry Golding. They commented that they had been coming to this basin during the summer, almost every weekend for twenty years, and that we were some of the first people they had ever run into. As it turns out, they were camped at Blue Lake too! We chatted for a while and found that Jack lived just through the block from me and worked for Granite School District for years, as I had done. We all remarked about how small the world is, said our so-longs, and the two men went on their way. Mark and I stood on the ridge talking about a potential future trip into Middle Basin while the boys tossed rocks over the steep escarpment that then landed in the talus far below; we were all having a great time, and it was about 2:00 p.m. As Mark and I were looking down the canyon toward the trailhead, some twelve miles away in Christmas Meadows (which leads to Middle Basin), Mark spotted some lightning activity beyond and off to the northwest! The storm appeared to be headed our direction, and Mark thought that we should leave before we became part of it; as he bent down to pick up some trash leftover from lunch, suddenly at that very moment, a bolt of lightning struck me in the back of my head!

Our boys, who were standing behind us and witnessed the event, later said it was like a dream. I was standing there, then suddenly a big white light was on my head, I went rigid, then there was a big explosion, then I went somersaulting through the air ten to fifteen feet, disappearing over the cliff. They were both in shock from what they just saw! The explosive power from the lightning bolt sent Mark reeling backward and rightward away from the cliff. The bolt exited through the top of my left foot and blew me a dozen feet horizontally and dropped me another ten feet or so down the huge north face of the Mt. Agassiz east ridge. As I fell, the left temple of my skull was penetrated by a jagged pointy rock sticking out of the cliff, arresting my fall. I was literally hanging from a rock, stuck in my left temple hundreds of feet off the ground. Back up on top of the cliff, when Mark (badly shaken by the sonic boom) became coherent, he asked the boys where I had gone, they pointed and said, "Over the cliff." When Mark looked over the edge and saw me, he couldn't believe his eyes. Mark believed I was dead; I was full of holes and wasn't breathing, near as he could tell. The lower half of my body was dark purple, and the upper half was ghostly and stone-white. The point of the rock I was hanging from had entered my brain and severely damaged my neck and spine too; it was all that was holding me to the cliff. Mark didn't want the boys to see me like this, so he instructed them to go down to our camp a half mile away and find the two men we met earlier and get help. In the meantime, while he was watching me and trying to decide what to do, I suddenly took a deep gasping breath—my first. He couldn't believe I was alive! Mark said that I started groaning and moving, so he had to get down this class 4 rock (which is steep and loose, often requiring a rope and a belay) and get to me before I dislodged myself and took the final plunge. When he finally arrived at my location, he realized just how dangerous our position was; the cliff was very precarious and sketchy, and he needed help! Mark said a prayer over me and asked God to provide him with the means to get me and him safely back up this steep ground.

Months later, in a private conversation, Mark told me that while were out on the cliff, there was a third presence; I told him that I sensed it too even though I was apparently gone or unconscious. While this part of my rescue was happening, the boys had found Jack Booth and Larry Golding, the two men we had met on the ridge earlier, and told them what had happened; they immediately sent several of their boys down to a lower lake in the basin where a ranger had been, he had a two-way radio! They found the ranger, who then summoned an ambulance to the highline trailhead eight or so miles away, but it crashed and rolled in route, so a Life Flight chopper was sent to evacuate me. Meanwhile, as Mark finished his prayer, he said I started grabbing like I was trying to climb. He knew that this was his answer to his prayer. As I rolled onto one side, he slid his body undermine as good as he could and created a makeshift fireman's carry (Utah Air Guard guys were taught to adapt), then hauled me back up to the top of the ridge where he administered first aid and treated me for shock. A short while later, Jack and Larry and our two boys were coming back to the ridge to help Mark when they ran into two men; both were paramedics that happen to be backpacking into Blue Lake, they joined in to help. The rescue party was instructed to bring me down to the lake were the chopper could more easily pick me up and transport me to a hospital. So the rescuers put me in my sleeping bag then rolled and gripped the edges to create a makeshift litter and got me down from the ridge to one of our tents where the two paramedics kept an eye on me until the Life Flight chopper showed up. I cannot imagine what was going through all those people's minds, especially Mark and the boys. What if we had all been struck by the lightning bolt including Jack and Larry? This story may have never been told.

I for one in many ways am very sorry that it has to be told, but since I survived the ordeal, I owe it to so many people to tell the tale because people need to know what the mistakes were that caused the drama to happen, and they need to know why I lived, and they need to see and understand just how very much we need one another, even right down to complete strangers, savvy!

The chopper left Blue Lake with just me and its crew, arriving at the University of Utah Burn Unit around 7:00 p.m. I was still unconscious—apparently in a coma. However, I remember looking up at a white ceiling as fluorescent lights whisked by. Then I saw three faces—David Rowberry, our bishop, was standing at the end of, I assume, a gurney, saying a prayer for me with his councilor brother Brown, who was standing by his left side. At some other point in time, my brother-in-law Doug, a county sheriff, was standing over me; he was in uniform. They were all amazed when I later brought this up and could even tell them where they were standing and even what they were wearing. It was agreed that I was quite unconscious at that time!

The burn unit called my house while I was being routed to the hospital; my fourteen-year-old-daughter Angela answered the phone yelling to my wife who was just heading out the door, "It's for you!" Teri was angry about being delayed as she was already late for a business meeting, but she answered the phone anyway!

The medical people on the phone said, "Mrs. Oliver?"

"Yes," answered Teri.

"Your husband is en route to the University of Utah Burn Center via Life Flight and will be there shortly. Your husband's been struck by lightning and his situation is critical."

Teri asked, "Is he alive?"

The person from the burn center said, "Sorry, we can't give that information."

Teri then asked if our little boy Eric was with me. Again, they had no information about the rest of my party except that she needed to come to the burn center right away. Teri nearly collapsed. When she regained herself, she alerted all our friends and family and told them what had happened and would keep everyone informed. Later that evening, she informed everyone that as soon as possible, they would allow me to have visitors, that it could very well be the last time they would see me alive! Mark and the boys made it out of the mountains the next day and came straight to the hospital to check on me (Don't forget, there were no cell phones back in those days!). I woke up in the U of U's burn unit about thirty hours after the event, with my wife standing at the foot of my bed; she was giving me the most detached, distant, deep, sorrowful stare.

Later, I realized that I put that look on her face all because I didn't follow two strong warnings: one being my own gut instinct—I knew better than to not heed these types of warnings! I was bandaged head to foot, and I was wearing a neck brace, I could see that things were probably not in my favor! I also don't recall seeing color as everything seemed black and white. I don't recall feeling emotion or any pain, and I felt as if only part of me was present and felt like I was in a tunnel!

So I asked my wife, "What happened?"

Teri told me that I had been struck by lightning.

I was, I said, trying to grasp this detail, I couldn't believe it! "Where was I when this happened?" I asked.

She then told me that I had been on a backpacking trip to Blue Lake with Mark and the boys.

I was stunned because I could not remember any of it! "Where are they at now?" I pleaded.

"They are on their way out of the mountains," she answered.

They had stayed an additional night up there after the event. Mark later told me that none of them got a wink of sleep, I responded with, "I am sure!" I was quiet for a moment, then asked what the outlook was. Teri said she wasn't supposed to tell me. "It's my life," I demanded, "it's my right to know!" She told me that the doctors thought I was dead right after the strike, but hitting the earth may have jump-started me, but that due to all the damage, I wasn't expected to live more than a few days! They told her that I might stick it out up to two weeks, but time would tell as there was just too much damage. The lightning bolt had entered the back of my head and exited out a two-by-ten-inch section of the top of my left foot and my sock, completely destroying a fine pair of Asolo backpacking boots! This was my major exit wound and was third degree in nature. The energy also eliminated 80 percent of both of my eardrums, doing major damage to my inner ears as well!

I had smaller exit wounds on the front of both hips and on the bottom of my right heel too! All of the toes on my left foot were broken as were most of the vertebrae in my neck plus several in my spine. I later learned while getting a tooth implant, that some of my teeth roots had welded to my jaw. The original tooth had to be broken away from my jaw in pieces—no joy! Everywhere there was a zipper in my hiking pants had a burn of some kind! I also had a four-inch-wide burn from the back of my head all the way down to my tailbone that was a mix of second- and third-degree burns! Overall, over 17 percent of the lower half of my body was third-degree burns, including cuts, contusions, etc.

The lightning had also had burned out all of my body fat. The day I got struck, I was 132 pounds and contained 6 percent body fat, now I weighed in at 117 pounds!

Eat your heart out, Jenny Craig—I lost all that weight in about two seconds—now who's the man?

We would soon learn that my motor functions were dysfunctional, and some were nonfunctional at all. All of a sudden I realized I wasn't who I use to be—the husband, the dad, the master sergeant, electrician, or the climber! Now I was something else, some mass with PTS (post-traumatic stress) waiting to be transitioned into the hereafter. Seeing my whole body bandaged complete with the neck brace was quite humbling along with the other outlooks, which was, that if I lived, there was this so-called Black List of things that was or could go wrong! The doctors had told my wife that in the event of me living, that I may never walk again, I could experience severe personality disorders. Basically, they were unsure just how affected my brain function was; there could be organ function questions, I could go into cardiac arrest at anytime, have seizures, etc. This list didn't make living sound like a very good option, and besides, they didn't think I was going to make it anyhow. But the IRS (Internal Revenue Service) has different ideas on these matters! Somehow I think they helped me survive as they hate to lose taxpayers.

Anyway, a day or so later, I started getting quite a few visitors and realized that the end could be very near! My oldest brother, who loves me like a mother, was sitting along the right side of my bed right next to my lifelong friend Mark Gomez; those two guys looked awful, just full of sadness. We talked for what seemed like a long while, and I remember thinking that they looked worse than I felt! Then Harry came by with a gift for me and demanded that I get better. About a week later, my classmates from college and our teacher Garth, who was also one of the electricians at the U of U Hospital, came by to wish me well. One of my classmates, Shaun, told me years later that they had been told that it was the last time they would see me alive!

So many people came by early on and wished me well and brought me cheery things like climbing books, but I couldn't read them because of drugs like morphine and my slow brain function. I just couldn't seem to focus or find the interest to read despite the awesome gestures! My brain seemed very inactive, almost without thought! I also received beautiful flower arrangements and Reese's Butter Cups, which I love but couldn't eat just yet. I received encouraging get-well cards, and all for a guy who was on the chopping block! I felt sorry for putting all these people, including the rescuers and the medical staff, through this trauma.

Time seemed to be, when I was awake, very slow; and when I was out, it slipped by quickly, like my life, I thought. But pretty soon, it had been three or four days, and I was still around, "like a good taxpayer," but living under a shroud and in a prison.

Around day 10 of my stay in the burn unit prison, I was sitting up in my bed alone in my darkened cell waiting for death, not wondering when it would happen, just waiting for it to happen. Later, while looking out my window, I saw a man jogging up the road that circles around from the north to get to the hospital; I remember thinking, *What a lucky guy you are to be out in the sun feeling its warmth, and I am stuck in here on death-freakin'-row*, and realizing that I finally had a thought that was attached to an emotion. He had no idea that a few hundred yards to his south was a place full of such high drama—of death and survival and unimaginable pain.

Burn units are full of this and more; lots of question marks, and I had just become one of them.

I began a prayer asking God to end my life! *Come and get me*, I challenged. *What's the point in dragging this situation any further? We are wasting time and money and bed space.* After all, the doctors apparently thought I was gonna go any day now! Suddenly, I remembered that I had a family! I also remembered something that I had told my son Eric, which was to never give up, never, so long as you have breath in you and a heartbeat! I had done my best to iron that idea into his little head, just like a karate instructor had done with our class years before. Now here I was throwing in the towel; I was giving up on my wife, son, and three daughters! I was willing to let life chew on them without me there to help! The dark side, I realized, was alive and present. That side likes to see these things happen, it's just one more way to break up families and get a toehold against the good in life; pain, misery, and defeatism is their calling card. My life had been about winning, now I was teetering toward losing by giving in.

I was taken aback at these new thoughts. Why hadn't I thought about my family even once till now? And here I was voluntarily placing my head in the guillotine and felt very ashamed. Even if I do die, I thought, it is my responsibility to fight death to the end and to do my best to provide for my family till that time has come and gone. So I told God that if there was any chance at all of a way out of this, I had to come back whole in every way (I always ask a lot). Then I ended my prayer to the man and realized just how close I come to willingly throwing it all away.

After several weeks I was still alive and had improved a little, so the doctors let me go home to be with my family and friends for a short while. I guess I was stable enough at the moment, and my wife had been taught how to care for me and of course was given direct phone numbers to medical help if things came to that; it also saved money.

One time while I was away from the burn unit, my guard unit had the base photographer take a picture of Mark and me together with the boot, sock, pants, and hat I was wearing the day I got hit. The clothing alone tells the story of something very powerful that went to work. The picture—well, I don't like it because I look like a person who is dead, but the body has just not realized it yet; I looked so transparent, it's not me. After a day or two of being in my normal environment, I had to return to the burn unit prison to receive some more professional care. This process would become a routine. I would receive needed care, then I would become stable and get to leave the burn unit for a few days, which was always welcome, but it always seemed to create anxiety because the freedom was short, and I would again find myself back in prison facing more questions about my outcome.

The survival roller coaster had begun. As Higher Powers would have it, change was again in the forecast!

One day, three or four weeks after my arrival at the burn unit, Dr. Saffel, my burn surgeon, came to me and told me, "Young man, you are supposed to be dead, you had too much damage, but you are alive. There is something you are supposed to accomplish while on earth, get it figured out and get it done as your time left is a question mark!" I felt a pang of emotion deep within me, so deep it could have been out in the universe. I seemed to be getting some emotional responses happening every so often; vastly important, closely held things in my life were still alive in me, so I was alive still. I have never forgotten what Dr. Saffel said that day while I sat there in a wheelchair holding my head up with my hands helping the neck brace support my shattered vertebrae.

I was still alive, and part of me even wanted to be alive now!

I will tell you this though: there was an internal war raging on the matter, and the survival roller coaster always seemed alarmingly close to the edge of coming off its track.

One day, a pretty blonde showed up in the burn unit to see me! She had come complete with a clipboard, perfect hair, a snow-white smock, a huge smile that showed off the biggest whitest teeth I have ever seen in a human mouth!

As she announced herself, I cut in saying, "I know what you are!"

She kinda condescendingly asked, "What am I then?"

I told her that she was obviously a psycho-cologist.

"How do you know that?" she quipped.

"Three things," I said. "One, you have the whitest smock in the place. Two, you have the whitest teeth around, and three, you are the only one smiling in here. Do you see anyone else smiling? So you must be a psycho-cologist!"

Giving me the quizzical look, she said that she had a number of questions to ask me, and my job was to give her a response to the questions! After she got started, I immediately recognized the routine: there were about twenty-six questions overall, covering several topics, and there happen to be about three questions relative to a certain topic asking the same thing, just in a different way! I thought, man, I am almost dead, and I recognize this trick, these therapists need a therapist! The military examined us this same way off and on, trying their best to figure us out, but we usually figured them out first. My reaction to this routine told me something about me: I was getting some of the old me back, even though there was still a big hill to climb.

The therapist began her routine by asking me if I felt like killing anyone, so to have some fun, I said yes; surprised, she asked who. I said, "My mother-in-law." She asked me why, so I replied with my most sincere face and tone, "It is because I think it was she that sent the lightning bolt to get me." This part of my therapy never seemed to do me much good anyway, and it lasted for about a year—eventually the questionnaires (very repetitive) were sent to my home where I would answer the dumb questions and mail it back complete with lengthy answers. I believe this should have told them they were not helping. Lucky me, eventually these questionnaires stopped coming. This work, I believe, is way too generic; these people never really got to know who I was or what made me tick, so how could they expect to do any good and help me? It was actually quite insulting to be treated like a serial number because burn victims' problems run very deep and can last forever physically and mentally! Surgery and physical therapy were happening too; I was being put back together. Overall, I knew things were improving because the food got better. Instead of Eggos and watered-down juice, I was put on a five-thousand-calorie-a-day diet consisting of real food that was very fattening, which means lots of the good stuff—that is, fried chicken, roast beef, and mashed potatoes, even trout and rocky road malts. So much for competing with Jenny Craig!

Something I learned about doctors while I was entrenched up at the U of U burn unit is that when they want to make a point to a patient/inmate, they send a really big nurse or six little ones.

In the case of my new diet, I will tell you, 5,000 calories a day is a lot of food. The first meal I got under this new process was breakfast. One morning (I guess), this huge nurse (probably Hulk Hogan's mom) shows up with this big tray of food, announcing, "Hello, Mr. Oliver, here's your breakfast."

I replied, "Well, who's eating with me?"

This big nurse then said, "It's all for you, and I am here to see that you eat all of it!" I choked the food down only because I was sure she could overpower me in a wrestling match, but honestly, a few hours later, I was starving. Around this time, I was about to get my first burn bath (that, I remember anyway)! Lori, my six-month pregnant nurse, had gotten me into this big stainless steel shower of an affair and was now getting in with me, and I was naked! So I faced the wall in one of the corners and held on tight to a stainless steel support bar, all the while standing on the ball of my right foot, which was the only part of my feet that could support weight. Lori ordered me to turn around. I told her I wouldn't, that she was trying to take advantage of me when I was weak. She repeated herself, saying, "Turn around." I stubbornly and flatly said, "I ain't doin' it!" She then told me that she was far stronger than I was, so I better do what she says. I realized that she was probably right and carefully turned to face my nasty nurse who then looked me up and down and declared, "Oh my, the women of the world have nothing to fear with you, do they?" My jaw dropped, and I thought, *Why, you battle-axe!* Everyone else including my wife just wailed, they thought that was pretty funny; these nurses really do know their business and are just trying to help.

Another time, my inner ears started getting a treatment that I think up to this point I had experienced only when I was unconscious. I don't recall any treatments relative to my ears thus far, and I would have remembered this one! One day while lying in my bed doing nothing and thinking probably less, this little nurse shows up. I mean she tipped the scales at one hundred pounds maybe, I mean, if she took her thermometer out of her pocket, her frail little body would lean the other way. She greeted me saying, “Good afternoon, Mr. Oliver.” I said hello back; she then said, “Why don’t we roll you onto your side. I have a little medicine I need to put in your ears.” So I worked with her to get me on my side and into a position where she could administer the medicine. A few seconds after she had given me the dose, it was as if my head was on fire, it burned and hurt like crazy for about three minutes or so, I couldn’t believe the pain! I did some yelling and gritted my teeth. The next time that my little nurse showed up, she was with five others all about her same size, and I knew why they had all come (the big nurse must’ve been unavailable). I told them that they wouldn’t get me without a battle, so we wrestled and yelled a bunch; I thought I was winning, then they had me, and I begged for mercy, but got the medicine instead and was again lit on fire inside my ear.

While in the hospital, I woke up quite often in different places, either on a gurney or in a wheelchair; apparently I was being tested or worked on in some fashion or another! One day I woke up in some waiting room sitting in a wheelchair when I noticed some old lady also sitting in a wheelchair staring at me and popping her false teeth at me. I really had no idea what I looked like; the facts are I was white as a ghost, paper-thin, and the only hair I had was a swatch standing straight up out of the crown of my skull. I looked like a freakin’ peacock! I asked this old goat what she was looking at. “You,” she snarled. “Oh yeah?” I mooed, “Well, I’m in no mood for you.” “Oh yeah,” she snapped again, “you’re one of those freaked-out punk rock skateboarders, aren’t ya!” Then she held up a clenched fist, and I said, “Come on, battle-axe,” but decided to run instead and got my fingers stuck in the wheelchair brakes; she was still yelling, “Punk!” at me as my nurse hurried me away, saying, “I can’t leave you anywhere.”

I continued to receive visitors often; this really went a long ways to help the effects of the roller coaster of survival. Thank God for friends! My Guard buddies showed up routinely and, on one occasion, told me that if I give up and die, they were gonna freakin' kill me! One of them insisted that I owed him five bucks; I insisted that I did not owe him five bucks and I would live just to prove him wrong, the yahoo. At one point during my stay at the burn unit and during one of the low points of the roller-coaster ride, an Air Guard recruiter named Steve Anspah from Nevada (a personal friend of mine) came to see me. Steve whispered into my good ear (the left one) that I wasn't who I used to be right now, but I had to come back because my family and my guard unit needed me! Steve also told me that I needed to find just one solid reason to live and focus on it, not to worry about two things because I wasn't strong enough for that yet. The reason to live, he said, had to be more important than any pain I would have to endure to live; it must carry me through all of it! He also reminded me that in my weakened state, I needed to stay close to Jesus, that I would need his protection from bad influences. I felt an emotional pang very deep within me again, like it was way out there in the universe, but I felt it; I wanted to cry because somebody needed me. I never saw Steve again; he died from a brain disease a short while later.

Another recruiter, Sgt. Kalakis, from my base in Salt Lake City, came to visit too. Ted told me that I was going through an experience that many others never will and that I can now relate to a whole bunch more of people than I could before! Ted went on to say that now it will be my job to tell the story of what happened to me, which will help others get through their challenges. This same recruiter also did me a big favor! He learned from the hospital staff that I wasn't being a very good patient, that I was very disagreeable at times, and that I wouldn't help them help me. Ted asked a friend and business partner of his,

Jack, if he wouldn't mind coming by the hospital when he was in town and teach me some survival skills and how to be a better patient. Ted let me know that I would get this visit but did not say why or when.

One night, about six or seven weeks into my on-again, off-again stay at the burn unit prison, an older man wearing a pretty nice suit walked into my room and said, "Hello, Layne, I'm Jack."

I followed through, saying, "Hello, Jack."

He then asked me how I was doing.

I said, "Just fine!"

Jack then said, "Well, that's just great, but you know what scares me about you?"

I guardedly asked, "What?"

"It's what you think about when you are all by yourself!"

I thought, *Wow, this guy knows me!* I had experienced days when I just wanted to pull the plug and thought that being alive sucked; the "getting better" process seemed so unrealistic, it all appeared to be such little steps at a time, which were full of what I was, a big fat question mark. Crap, I even had to succumb to browbeatings with psychologists—I could be a mental case and didn't even know it. As far as I was concerned, I had been robbed, my dreams stolen, my life changed, or should I say, what life! My life was still in question—and according to the Black List, always would be. Jack knew somehow that I would lie to him just like I did with the medical staff trying to help me, so he used the fourth-person approach on me. If I had known there was something beyond third person, I would have been more prepared. During our short and only time together, Jack asked me if I had ever heard of a book called *Psycho-Cybernetics* by Maxwell Maltz. I said yes, that I owned a copy but have yet to read it. He then simply suggested that I needed to read it (there's that *need* word again)! Then he asked me if I had ever heard of Og Mandino, who was an inspirational writer. I shook my head a little, saying

I had not. Jack suggested that I read Og Mandino's first book, *A Better Way to Live*, and then wrote them both down on a piece of paper for me so I wouldn't forget. Jack wished me well then left! Amazingly, Mr. Jack was not in my room all that long, but he left me with a strong feeling of "you need to do something to help yourself right now" that I have never forgot! I began reading *Psycho-Cybernetics* as much as my brain would allow and came to a place in the book that gut-punched me and slapped me around pretty good! This sentence talked about the responsibility of the patient to help the doctors and nurses get the patient better! I had been mouthing that I wanted to get better, but I wasn't doing much to help anybody or myself get me there! I chose to focus on all the negative probabilities and damage and dying as opposed to living and dealing with the issues as they arose. For some reason, hearing Dr. Maltz talk about me doing my part had a greater impact than anyone else had had; this single sentence caused me to change my attitude and outlook. From that moment, I decided I would start moving forward and help the medical people help me; there was light to see, and I could see some possibilities, and there was and always will be a reason to try! Now this does not mean that everything was going to be smooth sailing! You see, the roller coaster was running full tilt, and the low days were very hard to overcome, and the high days were few! This is why I said, "Thank God for family and friends and also thank God for God."

I lived in this burn unit with people who had tremendous injuries to the body and the mind, and if they lived, most likely they would always be different; they would forever have scars inside and out. In many ways, their problems were different even if they looked the same. There was a guy that had been completely buried by an entire load of hot asphalt dumped on him, another man survived a boiler explosion; these were 100 percent third-degree victims. We had another guy (an electrician) show up that simply turned on a large main circuit breaker that exploded;

the incredible arc flash third-degree burned the entire front of his body. His comment to me when we talked was "I used to be an electrician"! What blew me away is that some of the most damaged people cheered the rest of us on; they seemed to be the ones who were the most upbeat, smiling, and asking, "How are you doing?" There was this big blond guy who had lost an arm and a leg and had third-degree burns all over his body! I never saw him lost in despair; he was always making it around the burn unit, checking on his burn mates. I was bunked next to a Japanese man who had been severely burned and, like most, in serious life-threatening trouble. His wife would come to see him and would always stop by my bed first and say a prayer in Japanese before she would visit her husband! One day I asked her why. She said that she had learned that I had been struck by lightning, and in her homeland, a lightning-strike survivor is held in the highest of honor; so she was paying me respect and honoring my survival, how humbling is that! I told her that I was deeply touched by her act of kindness and respect and was very sorry that her loved one had to endure so much pain.

On occasion, I would get taken outside and pushed around in a wheelchair by one of my nurses; the sun felt so good, but people were always staring or looking at me out of the corner of their eye—I hated it, I hated being different and having to rely on so many people to get me through the days and nights!

One day I was being pushed along out in the sun lost in my little world hating my shadow on the walkway in front of me when this worker from the hospital came walking by. He was looking at me out of the corner of his eye when I blurted out his first name, Pat. He was one of my closest buddies in electrical school; we studied together, broke calculators together—you know—the ones that defy you by going blank at critical times trying to get you bad grades. Hell, I stabbed my TI-35 with a mechanical pencil one time for its acts against me, the bloody rotter. Once, Pat smashed his one hundred-dollar HP reverse polar against a block wall! Garth just shook his head; what could he do, we were freaks! Pat was stunned when he realized who I was; he hadn't come with our teacher and the other classmates before. "Layne," he cried, "oh my god, it's you." He said he was so sorry for not getting to me sooner! We talked for a while; then, he had to go, but before he left, he said, while hugging me, "You have to come back, I need you, I will never make it through school without you," then he was gone. As my nurse pushed me along, I noticed my shadow again on the concrete and was realizing how much I hated being reliant on so many people for my survival and mostly how much I hated being different, when for some reason, I remembered Larry.

Larry was a kid from my neighborhood when I was growing up in Kearns and was the target of others' ignorance! Larry was mentally retarded and was treated with cruelty by quite a few people. Routinely, Larry would come to play basketball with the rest of us, but we would run him off calling him retard, stupid, etc., making him cry, but Larry would go get a big softball-sized rock that, by the way, he could throw straighter and farther than anyone and start acting like he was gonna bean somebody, so he always wound up getting to play! Larry was retarded—and so, was different from the rest of us! We kids didn't think he felt anything; after all, he was a retard, right? Wrong! Sitting in this wheelchair—different now, a question mark—I got a small taste of what Larry went through in his life, and I cried; how could I be so stupid to think that Larry had no feelings? Crap, that poor kid felt everything we said and did to him and had to live with it behind the walls of his own little boxed-in world. God sure has some interesting ways of teaching me lessons!

The time had come to repair the damage to my left foot; a technician began debreeding and scrubbing my left foot, preparing it for skin grafting. The tech gave me a towel to bite (a substitute for painkiller), apologized for what was about to take place, and then began scrubbing the two-by-ten-inch exit wound with a stiff-bristled scrub brush.

I can't really describe what it felt like; the word *pain* was meaningless! I was biting the towel and digging my ribs to the point of bleeding, trying to find something else to focus on, but it was no use; I could not hide from what the doctors describe as something worse than childbirth. I had trained myself to manage pain while in the mountains, and one of those ways was to inflict some pain somewhere else on my person to take my focus off the injury so I could survive. The other process I did was to not take any pain medication for anything; I did this for the last ten years, and here I was in a situation where my mental strength was losing big-time. I was crying and digging to no avail. My wife, who was there watching, got up and walked out; she couldn't bear to see me suffering. Around the room were other burn victims watching this process take place; one of them, a 90 percent third-degree burn victim, attempted to make eye contact with me, and I hated how he looked; it reminded me of what I had become. Eventually our eyes met, and he was giving me the thumbs-up, then he said, "You have to make it for the rest of us, we need to see you make it and to walk again and to be whole again." His words cut through the pain I was experiencing—like butter—and humbled me to my knees; I now cried for them! Many of them would leave this place only because they died, some of the rest would never be the same, and a lucky few would continue on in some version of a normal fashion.

The operation I endured after this preparatory work was accomplished was designed to install a skin graft that would be taken from my upper left thigh. During the surgery, I had an allergic reaction to the anesthesia and could feel the surgical procedures that were taking place; while in this phase, my blood pressure had skyrocketed in the mid-200s, and the staff thought

I would have a stroke or go into cardiac arrest. I was unconscious for several days after this event; when I woke up, I groggily told my wife, who was sitting by my bed, I feel like I have been asleep for a week. She said it's been nearly three days then filled me in on the details of what took place in the operating room.

Nobody knew this, not even me, but something from that drama got remembered deep in my mind; from then on, every surgery I had brought with it lots of anxiety and panic, which was quite a terrible ordeal. I believed I would go to sleep and never wake up! So I started requesting Valium before I would go under the anesthesia, which seemed to help as the surgery got down to the wire, "You know the drill. 100, 99, 98, 97, 96…"

After this surgery happened, it was time to get up in a walker to see how everything was functioning. Several therapists helped me get positioned in the contraption, and after I found my balance, one of them asked me to take a step! I knew that my brain told my legs to do what was expected, but nothing happened; I tried again, but there was no response from my lower limbs! I couldn't believe it. *Damn it,* I thought, *I can't walk, so I can't climb, so I can't get to a bathroom and lose this freakin' sucking catheter*, which by the way, the day it was removed, my eyeballs and several other body parts almost came with it right out the same orifice! The nurse that did the removal acted like she had just conquered Rome or something, the little snot! With my legs not working though, I was stunned, this was a huge pill to swallow, this put a whole new twist on the direction my life may go or not. I know that many people thought that if I was faced with not being able to walk, my life was over. The therapists who worked with me said things will get better. They taught me to swing my hips; I did so with a huge lump in my throat, and this is how I started to walk again—one swing at a time!

Your legs and your feet—just think of everywhere you have been because you had the means to travel there under your own power. My god, I thought every part of the body has a unique and necessary purpose, a value you won't miss for a second until it is gone. I spent some time in a wheelchair, which gave me some ground zero realization time! As you know, my ears were badly damaged too, but by God's will, I can still hear well enough to enjoy the Beatles or my wife yelling at me or my grandkids asking all their endless questions. I can listen to a bubbling stream or a jet streaking along in the sky; all of it is so sweet and precious in its own way! Did you know that there are just as many great things to hear as there is to see, but realize that the ability to hear and see something beautiful is a gift every time?

The ability to walk through a forest and hear it and see it is a gift every time.

The ability to see someone you love, to hear their voice, then to walk up and hold them is eternally a gift every time, so obviously, touch is a gift too!

The ability to thank someone who has helped you in some way is indefinitely a privilege and an honor!

One day while I was on the down side of the roller coaster again, two eighteen-year-old boys had been brought into the physical therapy room! These young men had been involved in a gasoline fire out in Cheyenne, Wyoming; their bodies had been completely burned. Today they would begin their physical therapy! I was on a big low to the ground table where my therapist could work on my legs and was feeling less than happy to be around. I watched the therapist get one of them (the one who would live), started bench-pressing five pounds of weight using a flat bench (standard bench press style), then the therapist walked away! The therapists will walk away to see what you will do: will you keep going, or will you give up? It's one of the indicators of your desire to die or try. This kid was so covered in bandages all I could see were his eyes; this kid's eyes were full of determination as he pressed the five-pound bar, he wanted to live. Again, I was cut to my knees and humbled in his presence—I thought, *What are my problems compared to his?* If he lives, his physical injuries will always be seen, mine will be covered up. If I die and he lives, I will have it easy, he will have it very hard! I right then realized that it is harder to live than to die; dying is easy, all you gotta do is jump from a high bridge or walk out in front of a diesel, OD, or shoot yourself or knife yourself or be in here in this burn unit and just give up. Living takes effort; you have to want to live if you're gonna! Look at this kid in here all burned up and trying as hard as he can to win, to survive. From this young man, I learned one thing for sure: that the eyes are truly the window to the soul! Sadly, his partner died not long after.

My new skin graft had to be watched closely and handled carefully while it was adapting to its new location! Gangrene was serious a threat, and I went down that road for a short while. Skin grafts are so delicate and cannot "take" for any number of reasons where as you just might have to have the old one removed and begin the process all over again.

I don't mean to be gross, but my graft looked like a thick cut piece of bacon lying on the top of my left foot and the three smaller toes; it was all neatly stapled in place. The surgeons attach these grafts with dozens of staples; overall, it's an imposing sight. Interestingly, the surgeon's assistant's name was Rosanna Lanza, which I morphed into a *Saturday Night Live* character played by Gilda Radner, calling this technician Rosanna Rosanna Danna! My technician Rosanna told me that everyone called her that, and it ticked her off; she then reminded me that the one who put in the staples while I was under the anesthesia would be the one who removes them when I was wide-awake, and Rosanna Rosanna Danna might be less careful than Rosanna Lanza. I quickly became her humble servant.

The donor site for my left foot was the outer thigh of the left leg, encompassing an area of about three inches by fourteen inches! The donor site must also be covered and protected while a new layer(s) of skin grows, which is a very lengthy and touchy process. The cover that was used for me was called a Bio-Brain and is similar to Gore-Tex used in outer wear and is, again, held in place by lots of staples. Bio-Brains act as a protective dome over the site until the removed skin layers are replaced with healthy established layers; then the Bio-Brain is removed, and life is good—supposedly. In my case, the Bio-Brain attached itself to my donor site and had to be removed the hard way. The first thing the doctors tried was soaking it for dozens of minutes in warm water; eventually they just tore it off, just like you would pull a tablecloth out from under dishes and silverware on a table—sounds fun, eh? It wasn't.

So the next idea was to get my donor site to complete the rest of the healing process with simpler covers, and if it blistered or scabbed, these problems would be cleaned away, and this process would be started all over again. One of the big concerns was that there was always the chance of getting infection, which could complicate the healing constraints! Another reality in this part of the medical world is that sometimes the doctors have to apply skin grafts to stubborn donor sites, and mine was headed in that direction. I couldn't imagine being faced with getting a skin graft applied to this donor site, then starting this routine all over again for the new donor site. This whole idea was very depressing and seemed like it would never end; it was one more thing to keep me down, but I was hopeful. It's like I said somewhere before: when you are hurt, time seems to slow right down so it can get every bit of suffering out of you that is owed.

Eventually, I got lucky, and we were able, through several months of effort, to grow good skin layers at the original donor site! I will tell you though, if you want to see something that matches the word *delicate*, take a look at a newly formed first layer of human skin; it makes a butterfly wing seem tough. In life, there are just some things you can't—as a human being—force to be. Whether you like it or not! My other burns fortunately didn't require skin grafts, and all I can say is that I pity the poor people who have to have skin grafts at all, but what about those 100 percent third-degree burn victims. Oh man, I don't want to even think about it, and I lived with them, I watched them get scrubbed and heard their cries and the sad stories of who they used to be.

I also had to have skin grafts installed to replace my eardrums, which as you may recall were 80 percent gone! Apparently, skin three layers deep taken from just behind the ear makes for perfect eardrums—I personally like them, especially since they were crazy glued in place. One time I was telling my story to a church group and was talking about how my doctor shined the cone-shaped light in my right ear and had his other hand on the other side of my head by my left ear and was making animal silhouettes on the wall with the light going through my head, saying, "Yep, there's no problems in there, in fact, there is nothing in there at all!" One of the older ladies, looking confounded, said, "Really!" Her friends elbowed her, saying, "Gladys!" Everyone laughed. Then I told them about how this same doctor procured my new eardrum from behind my ear and then went looking for it on the floor after he thought he had dropped it, when in fact, he had crazy glued it to his finger. Good old Gladys chimed in again with "Oh my gosh." We all sang with laughter on that one; Gladys shook her forefinger at me and laughed too!

My eardrums were then and still are an issue, but are doing okay. In the beginning before I had eardrums, I heard things that no one else heard, it drove me crazy, I was constantly looking around for the source of some sound! I had and have ringing in my ears including some new tones that have recently developed, which sound to me like a stuck motor or a wire shorting to ground. Recently, I went all over my house looking for the electrical problem/noise; then I went outside and decided to sit in my truck with the doors shut and could still hear this low-bass hum and realized it was in my head. I felt like an idiot and now know that I have some tightness in my neck causing this noise. I also have pressure in my ears, it's like the feeling you get when you go to higher elevations, except mine is constant; and what I think is odd about it is this: my right ear skin graft is loose enough that it leaks and air can channel from the nose and throat somewhat freely and out this ear canal. The bottom line is it's like always being in a vacuum or soundproof room; I've wondered too if this could be the source of some decent headaches. I don't hear out of the right ear, so I rely on the left ear to get me through life, which requires me to focus on what I am trying to hear; background noise lends me to be almost deaf as I can't make out what is being said etc. This problem makes playing a guitar harder than it should be, and tuning my guitar is done with an electronic tuner, which is just fine, at least I can still play and enjoy my instrument. The good thing about effectively being deaf in my right ear is when my wife is yelling at me or wants something that I don't want to give her, that is the ear I tend to listen with, it works equally well with my leaders, bosses, what have you!

11

A Version of a Second Death

As I said earlier, somewhere around a month after my arrival in the burn unit, my doctors allowed me some liberty. I could go home for a day or two a week, it was a welcome change of scenery. My wife had been taught how to care for me and was given the necessary emergency numbers in case something went wrong. My assignment was to rest, watch TV, and eat—nothing else. At this time, morphine was coursing through my veins. Did you know that it doesn't take away all the pain? It doesn't. There had been times at the hospital that a nurse would come into my room answering the distress call to find me begging to be knocked out. So with this on the plate, it was stressful for me and my wife to be away from the burn unit prison. Morphine does however take away your sense of real time. One time while on leave from the burn unit prison, I had decided "I was going to up the Operations Tempo" of my "learning to walk" scenario by using crutches.

My older brother Jesse was a physical therapist at the VA hospital at the time, and I had convinced him that if he would supply me with a set of crutches, I could help teach myself to walk during my time away from the burn unit. Bear in mind, he didn't know I would be all alone for several hours each day while my wife was at work at the grade school near our home; yes, the whole thing was a dicey deal, and I hadn't been very honest with the medical staff when it came to my being away from the protection of the hospital. Eventually though, with practice, I could stand and move around pretty good (my opinion) on these sticks; so the next time I was at home, I would attempt going up and down the stairs that led from my front door to my front room then down the other set of stairs that leads to my basement. It was slow going and grueling because I was so weak and my balance was way off too! I reasoned that if I could get to my ski waxing table in our basement, I was gonna try waxing my skis! Finally I made it downstairs and to my waxing table; it took forever to get set up while trying to maintain my balance and focus, but I did it. I managed to get my skis cleaned and prepped then heated my iron and dripped a grundle of wax onto the base of one of the skis; for me, this alone was quite a tiring chore. Then, I began running the iron over the cooling wax, heating it into the base of the ski. I thought I was doing great; I had one hand under the ski checking for too much heat and the other running the waxing iron, with my armpits balancing my crutches. It was the equivalent of someone driving up the freeway under the influence of alcohol, texting with one hand, and eating a McDonald's Quarter Pounder with Cheese with the other! All of a sudden, I noticed the ski base delaminating! It didn't make sense; I had been checking the underside of the ski for excess heat as I ran the iron over its surface, but my ski was being ruined anyway—crap, crap, triple crapola! The fact is, I thought I was okay, but in reality, I was not okay. I was not capable of assessing myself or my condition—I should be firmly and securely planted on my couch, but I wasn't; instead, I was faced with some very strong realities that were all very depressing. I really needed to pay attention to my doctors, but drugs and thick heads had prevailed, and I paid the price, again.

Another time while alone, I decided to go for a longer crutch walk than usual—whoops, I forgot to tell you that I had been making my way outside and into the sun and real freedom; it felt great!

Remember now, on my good days when I felt like living, I was beginning to become my old independent self and was trying to exercise and regain any level of fitness that I could. So today, my goal would be to crutch my way down to 3500 south, a major east and west road in West Valley City that is downhill to the north of my home and about 200 yards away. Whether I like it or not, it was an exceedingly unrealistic and dangerous prospect for me—if you haven't guessed already, as a routine, I bite off more than I can chew.

It took a while, but when I arrived at the stop sign at the end of my street (my turnaround point), I was thoroughly exhausted and still had to repeat the effort going uphill; I felt so stupid. All of a sudden, and to my horror, I heard a dog growl! It was this white little mutt with a mouth full of teeth that routinely barked and growled and howled as it chased people and cars; it even nipped at airplanes when they flew by. Now, I was in its sight, and I couldn't run or defend myself; he had closed the distance in a flash and was biting through the bandages on my left foot, just tearing at it! Frantic, I spun wildly on the crutches yelling and screaming; there was no one there to help! I was so scared, I was sure I would get badly injured or worse. Somehow I managed to break away from this devil, and while standing there with gallons of adrenaline rushing through me, I realized for the first time that I really did want to live. I had fought the best I could for my life, and for what seemed like a very long time, in the end, I was bleeding, still a bit scared, but charged with energy. I seemed to be, at that moment, very clear in my mind too, which was something a bit new to me lately. If I really did not want to live, I would have given up and let this creepy dog have its way, but this battle spelled things out in its own pragmatic way.

The dog fortunately moved on, leaving me to drag my injured body back up this hill to my house several hundred yards away where I, with the adrenal valves closed off, collapsed on my couch, experiencing an emotional realization of who I was, who I am, and the long road ahead. I have to tell you that later when I could drive, the dog's day of retribution came—and by my hand and done in my own unique way!

One day I was driving my 1989 Toyota 4Runner, and as I turned the corner off 3500 south to go up my street, there was that miserable dog standing out in the middle of the road threatening someone. I shifted the 4Runner into low, gunned the engine, and lay on the horn, beelining right at the dog. He saw me coming, and his eyes popped right out of his head; scared out of his wits, he started running directly away from this runaway locomotive. I chased this mutt right up the road; then he bailed onto the sidewalk. I continued after him until I had to dodge a mailbox or two, the dog barking at me as I went by. But redemption—at least a little—was mine, and that is what that miserable mutt gets for causing me the fright and additional anguish of repairs to my skin graft area.

Eventually, I was well enough to be allowed, during my liberty from the burn unit, to go back to work for half a day several times a week—if I could handle it. Part of the reason for going back to work was a result of my wife informing me that after the several weeks of convalescing, my sick leave from my new job had run out, all twelve days' worth. What amazes me is the job I had left a year before so I could work at the Air Guard was a state-funded agency too, but would not transfer my sick leave to my new agency. I had left 950 hours of sick leave behind; it represented the fact that I had taken exactly one sick day in ten years of service, and now I was in trouble and needed this help and couldn't use it to help my family stay afloat. Now financial challenges would be building because a major portion of our income would not be coming in, and it was all because of me. I thought about how easy dying would have made things, for me that is, as the dead have very few responsibilities. As for the living, I also did not know that my wife had been confronted with other hard realities stemming from our insurance company and the hospitals concerning my accident. She had tried to keep me as much as possible away from the negatives that I couldn't do anything about so I could focus on healing my mind and my body. For example, Life Flight had been dropped in the July 1990 employee negotiations, so we were faced with a multi-thousand-dollar bill for that service. Also, there were hearings with my wife—the pilots of Life Flight versus the insurance company, who wanted to know why I was brought to a regional burn center—to save my life and treat my burns was the answer! But of course, the insurance company was doing everything they could to get out of paying the bills. The insurance company insisted that I be relocated to Holy Cross Hospital in downtown Salt Lake to have my ears repaired because they didn't feel that my ear damage had anything to do with the burns and so forth. Also, some of the floors or departments at the University of Utah medical center were not covered by our insurance at that time. Other issues were cropping up too that the insurance company wanted my wife to monitor—for instance, the charges for services and medication I was receiving. Apparently, there were several discrepancies, one of which was we were being double and triple charged for medication. I would receive a pain pill for—let's say a Monday morning, and it would in fact show up on the itemized bill two or three times. Well, okay, don't we have accounting departments for this! My wife's stress level just had to be out the roof, plus she was faced with the fact that I might not make it in one way or another, and any one of those probabilities would really beat her and my kids up pretty badly. I couldn't believe what one bad decision had caused thus far, and this game was far from over!

Since the doctors agreed to let me go to my workplace for a few hours a week, without their permission, I opted to stay all day! It was so hard though. I had to hold my head up with my hands to give my damaged neck a break; then, I would go back to work, and I was doing all of this in a wheelchair, which meant my wife would bring me to and from work, adding more stress to her day! My military leaders were cool; they agreed to let me stay as long as I didn't jeopardize them or me. I was allowed to work with our unit's supply office microfiche, looking up parts and other supplies. The microfiche was part of a large government network that was kept on film plates and viewed through a magnified viewfinder. It is tedious, and neck- and backbreaking; but no matter what, I had to do this, I had to help with money and build up my sick leave—and, I had to come back! I was very grateful to my colonel; since being allowed to stick around the base, even on a limited basis, increased my exposure to my natural world too, and for the most part, it felt good to be back and it gave me hope. I had one other dilemma that was riding my back like an ugly wart. When I had been hired, I had agreed to go back to college to increase my education and job skills as a journeyman electrician; our workplace had lots of special equipment most electricians never get to see, and I had to be able to manage it. As it turns out, my event occurred just before the beginning of fall quarter, and in the electrical apprenticeship program, if you miss three days or more in a quarter, you are out of the class. It simply is too technical and fast moving for you to hope to catch up! The agreement had been with the facility manager at my new job that they would pay for the school and books, and I would apply myself, get good grades, and graduate; if I didn't complete the course, I was out of a job. I went to the facility manager to talk with him about my options; he told me that there was no training money to keep me in school. My employment with the Guard seemed to be doomed. I told him that I was going to go to school anyway, but I don't think he believed I would or could make it. I simply could not lose my job too. These were my National Guard friends; this was my unit, which I had been part of for fifteen years. I had never said no to a deployment and always went on the missions that had the highest degree of need. I would get better and continue on like nothing ever happened, I just needed the opportunity and a little time and a little help; after all, they had hired me under an agreement. Fortunately, one of my recruiting friends helped me make some money in a commercial chemicals sales business he owned, which paid for tuition, books, etc. From this, I realized how true a statement I once heard was; it goes like this: In life, there are fair-weather friends, and there are all-weather friends!

My father-in-law, Howard, who at the time was in charge of the Utah state prison guards, became aware of my dilemma over the tuition and book issue, so he went to our HRO department director and found that there was plenty of money in our account for training, schools, etc. My facility manager was forced to pay me back for the tuition and books I had to purchase and continue with his agreement with me. Later he had a private meeting with me and made it clear that he did not want to pay me back and told me that I better never put him in that kind of spot again or I would be sorry! Crap, that freakin' idiot, I was already sorry and seemed to be getting sorrier at the hands of my military bunkies, or should I say flunkies! I was really too weak overall to debate him, but I felt the dog tearing at my foot again, and I was worried about my survival again. It was also agreed between the Air Guard and the burn unit doctors that getting me back into the uniform and into my military environment would be real good for me. So after reporting to the September drill—in a wheelchair, all bandaged up and half dead (more or less a visitor that bothered some people), I reported for October drill in uniform minus the left combat boot. I couldn't wear a combat boot on my left foot because it was inside what is called an Unna boot, which is a compression cast designed to protect skin grafts and control the scar tissue. I now was also using crutches (most of the time), which had been prescribed by the burn unit staff, not me.

While standing in opening formation during this drill, my first sergeant came up to me and flatly asked, "What are you doing here?" I produced a slip of paper from my breast pocket; it was a doctor's note releasing me to duty. I told him that the commander and our base doctor and the burn unit doctors decided that it would be good to get me back into my normal environment. The First Shirt (sergeant) examined the note then me, stuffed the note back into my shirt pocket, then grabbed me by the arm, pulled me out of ranks telling me to go home and not come back until I was ready. Fortunately, I was still pretty numb in the head or I may have responded and got myself in trouble, plus I couldn't go right home; my wife had brought me to the base because I was unable to drive at the time! So I crutched my way through the snow to the medical squadron a quarter mile away and told CMSgt. Foote what just took place—man, was he sore. The chief made a phone call to my commander and then put the wood to my "First Shirt." While the chief was on the phone, I heard a lot of "yes, sirs" and "no, sirs" and apologies; after he hung up, he told me he would give me a lift back to my unit. I said I was a little nervous about going back. The chief told me, "You'll be fine. I am here for you, that if anything else arises, I'm a phone call away."

Routinely, I would recall what the recruiter Steve Anspah whispered in my ear one dark night long ago: "You have to survive, your family and your Guard unit needs you!" Wow, it sure didn't seem that way with the Guard at the moment; I seemed to be a problem for some of them. I was definitely in survival mode, and it seemed that the longer I lived, the more things I had to survive against. I would come to learn that it was hard for everybody. But for me, life was up then down and often upside down! Being dead seemed, at times, better.

Few people at the Guard knew how to act around me; they didn't know how to talk to me or what to say, most people were worried that they would say or do something wrong. All of this is quite understandable, but the worst of it was that I recognized their guarded process when I was around them—crap, it sucks so bad being different. I wish people would just act normal, but they couldn't, their paradigms and fears wouldn't let them; some of my military friends would see me coming and do everything they could to avoid me—things were difficult on both sides of the fence! Even as I steadily moved into the realm of the living and away from the chopping block of the dying or dead, some people held the belief that I was not going to survive; to some, the question was, why waste time and money on a question mark? Man, it seemed that I was right in the middle of a second death. I seemed to have survived a lightning strike, now I had to survive its fallout. The wars for me seemed to have so many fronts, and well into the perimeter, right up in my face, and deep within my person, find a reason to live; I thought I had, but often wondered if I really believed it. Find a reason to live… I cried out "why!" I'm outnumbered a hundred to one, but then I realized that this wasn't all that unusual for me; this caused a spark of enthusiasm, and I kinda smiled.

I had taken lots of calculated risks in climbing that would freeze others right in their tracks. I went into the fire when everything was on the line. I did the same thing as an electrician, and in that realm became known as "gutter boy" because I insisted on taking care of the dirty dangerous work as opposed to having my buddies becoming exposed to these elements. I didn't understand it completely right then, but I would have to decide to live over and over throughout the rest of my life while wading through many different events; some would be straightforward life-threatening events, and some were just the muddy water of day-to-day living. I would also find change and adaptability to become fierce enemies, and I would have to learn to again embrace these ideas and what they challenged me with, or I was finished.

The Guard—throughout the next several years following the lightning accident—continued to complicate things for me! The idea of surviving a "second death" hung over me like a dark cloud.

One time while I was in the phase of wearing the Unna boot, a master sergeant who was part of our unit and my shop walked up to me and stood on the instep of the Unna boot and the barely covered toes, which had been broken and burned! As he pressed into my foot, he asked me if it hurt. I said no; another sergeant lashed at him, saying "What do you think, you idiot!" This guy was always bullying people, and I wasn't about to give in, even in my condition. This individual's act of stepping on my skin graft site could have easily complicated my life; fortunately, it didn't. However, in the meantime, my chief and facility manager decided that if I couldn't wear hard-toe work boots, they would have to lay me off at least as a civilian! So my chief took me down to base clothing issue and insisted I try to on some military "linesman" hard-toe boots; I protested that it was against the medical orders but was persuaded to try on these knee-high things that were, out of necessity, very stiff. Immediately my foot and leg was in pain, my skin graft had turned dark purple, and I was scared that I may have damaged the graft. As I had said before, skin grafts and donor sites need to be handled carefully or they could die and you get to start over. My donor site was not healed yet, and it had now been several months since it had been installed, but the graft was doing well, if we would leave it be, that is. Our facility manager (an obsessed micromanager) started demanding weekly medical updates on my condition and again had to be told that he would be kept in the loop as significant changes came about and that it probably wouldn't occur from week to week! Finally, the medical staff at the base and the burn unit had to get more involved and get control of my military and state leaders and force them to let me heal and move forward on a reasonable work program.

After I had gone through months of Unna booting and rebreaking all those toes one more time, I advanced to Jobst, which is a compression bandage and the final one. Jobst helps to keep the scarring to a reasonable limit and protects the area so it can safely and completely heal. Interestingly this was one part of my healing process that the insurance refused to pay for but is something that all third-degree victims need to undergo. You may wear them for a long time; I wore my pants and socks for a year.

I eventually would advance to a prescription leather work boot that was made out of glove leather, was very roomy, and had a protective toe that pretty much would allow me to be fully functional—under supervision of course.

Meanwhile, I had gotten back into school, and it was difficult, to say the least! I had to study topics over and over as my memory was not being very cooperative and my ability to understand the lessons was challenged! The feeling of being alone in the vast ocean of life was stifling and very humbling. So much for my solo climbing heroics, I thought.

But I was determined to succeed, that is, when I could be determined. But I had to force the issue because of the simple fact of this reality: if I fell too far behind, and this could be by simply missing three days in one quarter, I was finished with school for that period, and my job would then be in question, and my family's well-being would be at stake again too! I worked extra hard, locking myself in rooms for hours on end so I could focus and would receive little, if any, distractions. I managed to get an A each quarter that year, notwithstanding the topic that was AC theory, a very deep discipline that explores the science behind electricity. I guess my brain worked a little, and all along the way, I had little "bump steers" show up to give me reason to carry forward when I was about to throw in the towel! For example, early on during fall quarter, my schoolmate Pat grabbed me by the arm, looked me straight in the eye, and said, "Layne, you ain't who you used to be yet, but you gotta come back, I need you to help me get through this course." Again, I felt a little pang of emotion deep within me, and somewhere in there, I think I was crying because of all that had happened, and now I couldn't let go, I had to live. Regularly, more and more reasons came along saying I had to live no matter what it cost in money or pain; it's just the way it was. Pat and I worked closely together, we even broke some more pencils and calculators right along with the rest of our classmates—we laughed, passed tests, and swore to the gods that we would make it.

I can tell you, without people, even the ones that cared only a little, I would have been lost. All I needed was for people to say hello and look me right in the eye while saying it. I can't get over how understanding our teacher Garth was; he was so knowledgeable and patient, helped me succeed in and survive a very intense electrical program, and helped direct me in my electrical career. Isn't it amazing how people just show up with such perfect timing to help you move forward, then after they interject their brand of help, they are gone? I've had this experience routinely since my lightning strike. Interestingly, some of these angels are still hanging around. These days, I can say for sure, my family and I were lucky to have them come into our life.

I eventually learned from my time in the burn unit that how we adjust to life and its manners is a matter of perspective! That is, I was hurt very badly, but I could have been hurt worse, much worse, and still could have lived to suffer it out in some dark corner of my personal mental ward. Instead, I was back in school—it was a hill to climb to be sure, and I was receiving much needed assistance so I could move forward toward a new life that would be mostly good. I was so lucky, or should I say being watched over, even if I didn't fully realize it at the time. As winter approached, I was spending more days at work, which helped with money etc., plus it supported my small little gambit—that is, going to school and keeping longer days at work than I was supposed to, I was battling a fair amount of pain, which would become a day-to-day routine for the rest of my life! Please forgive me because this is easier said than done, but you have to work through your challenges, no matter how much or where it hurts, it's just the way things are.

My wife was also doing some concealing of her own pain; one version was in the form of intense stress. For one, she had a family to raise, but I think she was very concerned about the Black List of things that could go wrong with me too! I think more so because I had been pushing things a bit more than I should, and she knew I would eventually step it up exponentially trying to become who I used to be. I was also getting angrier by the day, and I was constantly finding new things to be mad about in my evolving world even though I was slowly but surely getting better. Many burn victims go through an angry stage that can last months and months; they experience all sorts of things, including intense pain and loneliness. If I had stopped to think about it and could be reasonable, I wasn't alone in my struggle (only in my mind); others struggled right along with me with their own brand of worries and much of it about me. The psychology that has always worked on me is to get me sore enough (mad) to do something positive and take affirmative action to right the wrong. I will at some point do the right thing, but sometimes though, I have to hit rock bottom and get humbled first.

During this point in my recovery, I was still on the 5,000 calories-a-day intake drinking lots of Ensure. My bad cholesterol was beginning to rise to alarming levels; it seemed that the good that was being done for me was also, at times, conspiring against me, which made me mad! Crud, two days before the lightning strike, my HDL (high-density lipoprotein) and LDL (low-density lipoprotein) were picture-perfect; I had world-class oxygen uptake and was fit enough to wrestle a bear, and I was gonna climb Everest.

Now, I simply became angry with everyone and at everything. I continued to view things from the perspective that I had been robbed, and this would cause me to be, at times, difficult to be around! My dreams were shattered, and surviving this hurdle was hard; what I didn't know was that it would get harder still as my body broke down later in life. Part of me was trying to survive and work through the issues because of my sense of duty and the reason to live that Steve had me focus on, but then the other part of me was not surviving very well because of anger over all the losses and uphills. It was so overwhelming and draining, it did nobody any good. Was I a failure? Or was I just hanging on to a fragment of feeling sorry for myself? I was still mad at myself for not being stronger during my scrubbings because I had trained so hard to endure pain, yet this injury caused me to become so ready to give up. However, I still possessed a well-known trait of mine from before the accident. As I said before: when things got real hard, I would get mad at my situation, and then I would want to give up; after hitting bottom, it was then that I would get another bump steer that would change my way of seeing things. I would get mad again and charge like an angry bull, fighting for my existence, so to speak.

I think this was one of the Creator's tricks in dealing with my perspective on life at this point in time. This cat and mouse game would be played for a long while, and it may have spared my life. In any case, somebody knew how to light my fire. Many good things had come my way, and there was more to come, if I could just keep myself together. Remember, the patient, student, or citizen has to do their part if they want to survive in life. Happily, my classmates and I all made it through fall quarter and was that much closer to graduating from our course. Christmas that year was lean; my wife and I were able to get the kids a few things but wrapped up items that we personally already owned so as to not let our children see that we were too poor to get each other a gift. This was a little embarrassing; but in the end, I was still alive, and we had each other, and this is what we wound up telling our children—this was our blessing and gift this year.

One day during January 1991 while at home on a rest day from the hospital and/or work, I realized I could fit my left foot with its Unna boot into my Koflach ice-climbing boot if I removed its inner boot! This type of climbing boot has an inner and outer boot that can be separated so you can wear the inner insulating boot inside a sleeping bag for added warmth, or to dry them out and keep them from freezing. So I got myself dressed in my winter clothes, grabbed my ice-climbing gear, got into my truck, and drove to a local ice climb known as the Great White Icicle, which is located in Little Cottonwood Canyon in the southeast corner of Salt Lake Valley. As I drove into the canyon, I got that familiar exhilarating feeling and sense of awe I always get when I see the towering granite rock formations, it felt so good! I was now in a place where I fit, I was a climber, and this was my home, and it felt good to come home. As I drove up the canyon, I realized just how much of my life had been spent in this incredible place. I also realized that I was breaking promises and rules by being in an automobile, let alone going to the icicle to climb some of it. I didn't care, I had had it with stagnation and the four walls of prison; to me, that was not doing me much good—and coming here was. Just being in the canyon was a real shot in the arm and felt like just what I needed. However, some of my thinking was not rational, simply driving my truck was a challenge, and the idea of climbing ice with a fairly new skin graft was not a real smart choice, rational or not. As I neared the parking lot for the climb, the towering flow of ice came into view, all seven hundred or so cascading feet of it. *Wow*, I thought, *holy cow, I have climbed that thing before.* It looked stunning with all of its angles and shapes, apparently frozen but ever changing with time, like me.

Actually, to a skilled ice climber, this is one of the more moderate ice climbs around, that is, when it's in good condition. The icicle is a cascading waterfall that looks imposing and worse than it actually is. It does have several near vertical sections and usually gets climbed in about five rope pitches and is quite fun. In bad conditions, it can become a real challenge and quite threatening. Anyway I pulled into the parking lot, which was devoid of other cars. I put on my crampons, my helmet, and then my pack, grabbed my trekking poles to use as makeshift crutches, and set off. Soon, I crossed Little Cottonwood Creek. After climbing a short grade, I arrived at the trail that runs the length of the canyon bottom and could once again see the upper two-thirds of the massive white icicle. *Wow, this is great*, I thought, my heart beating with excitement!

Despite being just a little lightheaded and winded, I decided I would pace myself and enjoy the several-hundred-yard hike up to the climb; but in the back of my mind, I was worrying about complicating things for myself by just being here, I was experiencing some very interesting emotions, and my head was now spinning with them. I felt like a prisoner of war escapee that knew he would be caught, but had just got his first real taste of freedom and good food in a long while, yet wondered if he would ever have it again. Prison sucks no matter what kind it is.

After a challenging hike up to the climb, I was fascinated and overjoyed to find the first pitch, which is about forty feet high in pretty good shape; it was totally covered in a six-inch or so thick convoluted sheet of ice with lots of rugosities, humps, and dishes. So I just sat down in the snow and took in the scenery for a while. Across the canyon (north) on the Black Peeler Buttress, I noticed toward its east end, the terraces that rise upward in a dozen or so cascades, ending near its summit—all of these cascades had either streaks of ice coming down from their lip or what appeared as columns of ice; it was a beautiful contrast with the surrounding granite and the vertical black streaks in the rock with the sun shining all around and a perfect brilliant blue sky. God, I thought, and I am alive. The smell of the pine was waking up and talking to my senses. To me right then, with the water under the ice percolating its way down to Little Cottonwood Creek, the towering cliffs above, my environment seemed very surreal. After a little while, I felt like I needed to grab my ice tools and go commune with the ice. So I tied a 10-mm rope on my back, slipped my gloved hands through the leashes on my Grivel Super-Courmayeur ice tools, then picked my way "down to the center" of the icefall, and stood there planting my tools and getting the feel and the sound and ultimately the taste of my expression. Finally I kicked in the front points of my foot fangs and moved up and down a few feet, testing the medium, and then I started upward. It seemed a little awkward to me at first, but I seemed to be instinctively doing what I should be doing, moving slowly and deliberately toward my goal—the top of this one section of ice. I could plainly see the rappel anchors on the giant stone to the right of this pitch and continued upward. As I climbed, I was learning about myself and was making some decisions about the war I would wage against what happened to me; and in the end knew full well that this war would be me against me—a very scary proposition for sure. I was nervous a time or two as I ascended this short section and never even noticed my throbbing left foot. I also noticed my sewing-machine calves (quivering calve muscles) several times, which I corrected by lowering my heels and breathing. Finally I made the final moves to surmount the rounded top of this section then clipped myself with some long runners to the rappel anchor and sat down in the snow to think things over. As I gazed around the canyon, I thought about how good it felt to be here and how good I did coming up the ice just now and realized how important it was to build a huge basket of reasons to survive and rebuild me as a whole; especially now that I was becoming more capable, I could battle the dark side of things, which seemed to be a constant threat. Really, I should not have been there, but I'm glad I was—to die there would be good, I was home.

Before I had been struck by the lightning bolt, I had been a pretty fair skier. At the end of the previous season, I had my Rossignol FM195s all tuned up for the next round. So toward the end of March of 1991, eight months after my life lesson, I snuck up to Brighton Ski Resort, which is located near the top of Big Cottonwood Canyon, just one canyon to the north of the Great White Icicle! I was there on the premise of checking my abilities, including balance, strength, and so on. But really, I was breaking back out of prison and ignoring the directions of those who were trying to help me. I barely even made it off the ski lift; then, during one terrifying run down Majestic's face (an advanced run that said it all), I was missing critical balance, coordination, and strength requirements. I could not believe how poorly I had done; I barely could maneuver my skis, and it was very hard to maintain even the snowplow position; it was depressing! So I went over to the Mary's run for total beginners and didn't do any better and blamed everything else for the struggle (i.e., the ski tech did something wrong in the tuning etc.). After several exasperating attempts, I gave up and handed my all-day pass to some kid in Brighton's parking lot and sold my skis to a National Guard buddy. How could climbing seem so possible and skiing not? My dreams and rationalizations just became a little bit blurred by a reality check—dang it, dang it, dang it. What was clear was that I still had issues whether I liked it or not and needed to be careful, very careful, or I might mess up and really change my course again. I was not paying attention to the big picture, which is not a good thing, especially when it comes to really important decisions and their attending actions and the associated fallout. Independence, my friend, is a good thing, but it can turn on you; it is always and forever will be a double-edged sword, and you must very accurately understand this and know how and when to apply it.

Later that spring, my confidence mended with a little time and with my curiosity nagging. I tackled something a little more involved and dangerous. I found that as I healed, I could get my left foot into my old oversized Boreal Rock shoes and figured that I could go do a rock climb; there was one that I had repeated many times and had guided others up: a very moderate 5.6-rated climb on the sun facing northside of Little Cottonwood Canyon called the Waltz! So once again, I snuck out with some rappelling gear, my climbing shoes, and chalk bag; drove to Little Cottonwood Canyon; laced up my rock shoes, then snuck my way up a technical and obscure approach route. After arriving at the base of the Waltz, I eyed the climb for familiar features. The climb begins with a lie back off a flake then moves into a steep ramp, which you will also use the lie-back process for, which by the way is pulling with fingers and pushing with feet, or in other words, creating opposition. Next, you reach a point where you directly approach a big off-width crack that takes you to the top of this climb, which in all is about half a rope length or a hundred feet, or you can opt to move out onto an exposed face and work around an awkward section at the beginning of the off-width crack. Either way, it's a very fun climb. Understand, however, that even though this climb is very low on the rating scale and is for the beginner, an unprotected fall from this climb could amount to several hundred feet, and you will most likely not survive it and will regret it if you do.

With the sun shining brightly, I worked up and down the first lie-back section a few times, nervous as a cat getting a feel for things. Finally, I drew in some air and began working my way up the route. As I neared the end of the second full lie-back section, I started to panic; the little bit of exposure and my weakness was all working to make me sweat like crazy, and I thought that I might freak out and let go. My adrenal glands were working overtime, so I backed down and thought about going home, but I was up here to see what I had left in me, and if I went home now, I knew I was finished as a climber and maybe with many other things too. I was really worried about my rational and motives too—like skeletons in a closet. But one thing I know for sure is if you're gonna do something important, be clear about how come, and then approach the issue with total conviction and be smart, eyes and ears open, and brain on / pilot light—'cause it's game on!

So I took a couple of deep breaths then worked my way back up the rock. As I entered the area of the lie back that turned me away, I had to suppress panic again and work the plan. I knew that if I backed down now, I should just go home and sell all my climbing gear and start up a hobby, like drinking to help me forget who I used to be. I quickly found some inner strength and worked through the problem, making it to a rest stance. After a short breather, I then executed a committing move out onto the dicey thin exposed 5.7 face. Three moves later, I was back into the obstinate off-width crack complete with the one-and-a-half-foot diameter, two-hundred-pound chalkstone that had rolled and threatened to trap me in the crack or pitch me off the mountain years before.

This chock stone is quite similar to the one that trapped a hiker in a slot canyon several years ago in southern Utah, causing him to amputate his right hand to save his life. (Check out his book, *Between a Rock and a Hard Place.*)

That day long ago when the chock stone had rolled on me, I was soloing just like today. I had worked up the ornery off-width crack when presently, I was faced with surmounting a big chock stone jammed into this steep crack. I checked it over pretty good by pounding on it and listening for hollow sounds or vibrations; I tried rocking it in several directions, it seemed sufficiently jammed. So I carefully began working my way over the stone, when all of a sudden, without warning, it began to rotate; as it did, my feet began skidding off their holds while the rest of me scrambled to not become trapped or get spit into a free fall. As quickly as it began, the rock stopped moving, rewedging into a new position. During the drama, all was electric, but when the big stone stopped, I couldn't believe how quiet everything seemed to be; it's as if everything around me saw the reaper coming and disappeared. Then afterward, I heard the forest, then me start breathing again.

Today, reminiscing about that day long ago, I was sitting at the belay station at the top of the climb thinking about how lucky I had been I thought about how I fought to stay connected to the rock, to stay alive; the moment was very intense, and I am sure it was full of emotions that I didn't have time to recognize. I was forced to act right now and adapt to the change lest the whirlwind take me where it will.

As I sat there contemplating the universe, I realized that despite all the bad things that came with the lightning, good had come too. This idea had crossed my mind before, and I probably will bore you with it a time or two more, but here I was in a place that I simply loved to be in; birds skirted along the rock, hovering on the updrafts in a crystalline-blue sky. A lizard scurried by, making my climbing skills seem novice at best. I really wasn't trying to be reckless today, I was trying to find me, I needed to know some things: who I still was (this question made me cry) and what I was still capable of doing despite the pain and physical limitations with my spine, circulation, foot damage, and brain trauma—that's a hot one to be counseling about three to four hundred feet off the deck. I wondered if I was still mentally capable of really being able to operate safely in the vertical world. All today proved is that I want to be here, not that I should be.

I tell ya, with me, my dreams die very hard; I am a dreamer not only at night but also in the light of day where I can get my will to take the dream and convert it into a vision and then into the tangible. So when a dream is stolen or is lost, I take it very personal; at times, it even seems like a matter of life and death.

Far below, I could see the beautiful forest full of all kinds of vegetation and Little Cottonwood Creek as it winds its way so free, between the glacially carved granite walls to make its way out into the valley and eventually into the Great Salt Lake—so beautiful. I am so lucky to have the eyes to see it, a nose to smell it, to be able to hear and touch it, so dang lucky. How could I leave this place, its allure is so overwhelming.

As I sat in my safe little nest high on the Gate Buttress, I thought, *I must continue this part of my life somehow. I can do it if I want to. Get rid of fear. Mind over matter, that's all it is. If I allow myself to get better despite my limitations, I can live here again in Little Cottonwood Canyon.*

12

School

Getting back into school had given me something else to focus on, something to make my brain work and grow in a positive direction. Reinvestigating myself as a climber and skier had given me something familiar to identify with; somewhere I had been before something that made me feel good, even if skiing seemed beyond the scope of possibilities, at least for now. Thinking back to a year ago, 1989 that is, I was one healthy son of a gun; I climbed rock and ice routes, humped big packs all over the hills, prepping myself to go to really big mountains. Now, I would be lucky to just get to a quarter of my original condition, and that's if I trained extra hard and for who knows how long and to what outcome. But I kept remembering where I had just come from, and if I think I have it bad, what about those poor souls that I left behind in the burn unit prison? It was hard to forget about them and that place. All they wanted was to see me be whole again, to walk again, to be normal. I had, to this point, survived an epic situation with Mother Nature, and I wasn't happy with my surviving the event, I am so selfish…

I went back to the burn unit one day during late March of 1991 for some more evaluations and work; it had been almost seven whole months since my initial incarceration. While I was there, I went to the therapy room. When I walked in, the ones that knew me cheered for me, and I cheered for them; the hope they had given to me a few months before, I now gave back to them! I was walking on my own two feet with my own legs with no assistance, so I did my version of Chubby Checker's twist for them; they laughed and clapped along. I could tell this was a real shot in the arm for many of them because they got to see that they too had possibilities! I really had nothing to complain about; I didn't live in that place anymore, seeing the destruction and hearing the agony. I was much freer than they, but like many of them, I was still locked into my own personal prison, internally at war with my past and my new world. I wondered how many of them would get to see their new world; would it be a good one? I wondered how they would adapt or adapt to adapting, or could they even get that far? Could they? How could they, did they have anything left inside to try? I had moved on to a new level, I was in that place only part of the time now, I was almost a human again, but part of me sure didn't feel like it, I thought. I had that war inside me that gave the nagging perspective of hating myself, and that part of life sucked.

My life now in many ways seemed to be racing away from the point in time of the accident; I was moving forward and past it. Something was keeping me alive and steering me toward the good, but it was a rough, rough, road; I was trying to understand it all… Like I said before, good things were coming; and as I analyzed things, I realized that I was becoming more optimistic whether I liked it or not. At times I felt like two people, but I knew that eventually I would become one, but I was cautiously interested in which one that would be. It was an uncomfortable period. My poor wife, Teri, had waded through and stressed over so many problems with the doctors and insurance companies, I can't believe she didn't develop ulcers; again, my problems seemed trivial to hers! I'll say it again I was on the road to easy street, dying or quitting or giving up is sooo easy and can even seem rational; you just make that decision and blow away with the wind. On the other hand, people who want to live have to work at it to stay alive; life requires effort and often is painful. This is the nature of life and human relations; we have to go to work, rotate tires, pay bills, stand in line at the DMV (Department of Motor Vehicles) waiting to see that ornery assistant who doesn't want to help us anyway; we argue with each other, have road rage, and even love each other once in a while—all of this requires regular effort, just like breathing. My wife alone had to keep our household together, even at this point in time, as I had to focus on college and continue to heal and get better and more stable physically and mentally. She was the one worrying and maintaining daily life needs for me and our children; I am not sure what she did to take care of herself. I really may never know the true depth and width of all her challenges; she kept me out of and away from all of the negative fallout so I could keep focusing on my return to a normal life in as smooth of way as was possible. How do you repay that level of dedication? The answer comes from Christ and should be in your heart.

School. Wow, it really did give me something to focus on. It really worked me hard; my memory had been affected by the lightning strike, which caused me to study things over and over to get them to stick, and sometimes that wasn't enough. My neck, which was severely damaged from the fall, caused me lots of frustration, making life difficult, and it continues to this day. While in the burn unit, I remember telling my wife that if I could get all the different pains I was experiencing down to a manageable number, I might feel more like sticking around. Interestingly when that number got down to just a handful, it turned out that these issues would be the ones that I would have to learn to manage and live with for the rest of my days; plus at the time, I was still confined to a wheelchair or the walker. Now I got another reason to get mad.

My wife, Teri, had to come and get me from work; then, she'd take me to school and would help me manage the wheelchair at each location. Fortunately after several months, I was up on my own two feet eliminating this problem for both of us. However, my neck, as I mentioned earlier, was the big challenge. It was so weak that after holding my head up all day at work (despite the neck brace), then for an additional four hours of school three nights a week and wading through all the homework—I don't know how I got through it, maybe by the grace of God.

Along the way, however, a new problem developed with my neck. While on the job, I might be working on something overhead like a light fixture and would need to tip my head back to get a better view; suddenly I would lose my eyesight, then I would feel like I was spinning then lose consciousness, oftentimes falling and sometimes from high places. So everything wasn't fixed, crap, what next, I thought! Now, my boss and friend Mark had to watch me closely (for my own good) to see how much risk I might pose. Again, I was worried about me in relation to my employer and where this may lead; really, I just wanted to be better and was trying to put my best foot forward to get there. I soon realized that this issue, if it were something permanent, could really dig deep into my life, including some of the dreams I was beginning to have about living.

Electricians and climbers articulate their necks all the time; it's just a matter of course. I simply learned to adjust the way I used my head to get a better visual on things (i.e., I would bend at the waist more than I would bend my neck and used my eyes more too).

I also believed that I could fix everything if I kept up with regular training in the gym and used physical fitness as one of my goals to help me survive; if I could keep heading toward a goal and stay fit, everything should turn out okay. I had to believe in something, or I might succumb to those thoughts of giving up, which is the road to nowhere. Unfortunately however, I eventually experienced a more severe event as a result of my neck issue.

One day I was boring a one-inch hole with a large Roto Hammer through a block wall in our paint-storage building on base. I was standing on a stepladder about six feet off the ground in a corner near the ceiling. Because of logistics, I was running the drill from its right-hand side with my head right next to the motor housing. While I was working, I must have looked up for a second and passed out binding the drill bit in the process. The drill motor whipped hitting me in the left jaw, fracturing it and knocking me six feet to the concrete floor. I awoke being attended to by medics.

After that episode, which was accompanied by a list of other close calls, my commander insisted on me being reevaluated. I was sent to a neurologist—who was supposed to be top of his class—to get his input on my situation. While visiting this doctor, we talked about what had been going on. While reading from some manual, he had me walk a straight line with my eyes shut; he also had me touch my thumbs to my nose—again with my eyes shut. Just like Kris Kringle in *Miracle on 34th Street*, I accomplished these tasks no problem. Afterward, the doctor, relying on his manual, said that I had a touch of vertigo and that my problem was all in my head (you know, psychological) and that people like me who had experienced trauma, often imagine that they have various other problems. Hmm, so now I am a nut, well, that's just great! Our base physician, a colonel, was sent the prognosis from this neurologist; he thought that it was time to end my career with the Utah Air Guard. Holy cow, this could mean both my military and civilian career; this would be devastating especially to my family. I had to fight for my existence. I argued my case. I pointed out that the eminent neurologist didn't listen to me nor did he pay attention to the details of my injury; he never really looked for the actual problem. This medical professional gave me several ridiculous tests then decided that I was a mental liability! It appeared that my career, and life as I knew it, was ending because of what someone with better things to do had read out of some book. The whole situation was insulting at the minimum; I felt like I was being swept away like so much dirt!

I begged the colonel to get another opinion! Another MRI revealed that I had some nerves being pinched around the C4 area along with some blood vessels and nerves being impinged closer to the base of my neck around the C7 vertebrae. It was decided that physical therapy could be used to help overcome my issues, and I was allowed to stay in the military for now. The thought of being kicked out of the military on a medical, for me, was a serious step backward. The Guard was, besides being part of my family, a major anchor in my life! I felt safe being a part of the Guard, protected if you will. It also gave me one more reason to exist, because I belonged and was needed there. However, this scenario drove home a reality lesson; I must stay in incredible shape, pay better attention to how I approach everything I do, and never complain about my physical and mental struggles—I must suffer to survive.

During the springtime of 1991, my family and I had been invited by Teri's friend Debbie to go swimming at a pool up in Park City. While there enjoying the day, I dove into the deep end of the pool; suddenly I felt a piercing pain inside my right ear. The pressure from the dive had caused my new tympanic membrane to come loose! You may recall, 80 percent of my eardrums had been destroyed by the lightning bolt, so the medical folks at Holy Cross Hospital used skin three layers deep taken from behind my right ear to make a graft suitable for eardrums! This membrane had been glued in place, and since I now tasted chlorine, I accurately assumed my eardrum was missing or had come loose.

I scheduled an appointment with an ear, nose, and throat specialist who also looked and acted just like Mr. Rogers and was the doctor who had installed the new membrane. I explained what I was doing and what had happened. Then I told him that I could plug my nose, pressurize my mouth, and cause air to exhaust out my right ear; his reply was, "No, no, you've had a terrible trauma very recently, and now you think you have all these other issues, it's really quite normal for this to happen, but it's really just in your mind." The test this doctor gave me was an inward pressure test effectively closing the canal with the flapping eardrum. I should have been given an outward pressure test too since the membrane was acting like a flapper valve opening toward the outside. This would have become evident and saved me the insult and a trip to another doctor.

So I went to a different ear specialist who was wearing a white lab coat with exotic pheasants embroidered all over it! Seeing his getup, I asked him to just listen to my story then we'd talk! So I gave him my case history and then told him about the swimming pool event, then the results of my visit with Mr. Rogers. When I was done, he gave me a suction test on my right ear, confirming I had a pretty loose eardrum.

Amazing, an ear doctor who doesn't listen, and one who does! I wasn't crazy after all, the horse's patoots!

The evolution of my injuries was a real hands-on bit of schooling! Very few of my injuries would ever be totally fixed; they would just be made kinda stable for the time being. My feet were also becoming a challenge! As I had morphed back to walking and being more active, the scar tissue associated with my feet was aggravating nerves that would eventually become neuromas that became tumors, which can develop into cancer. Neuromas are aggravated nerves in the metatarsal area that can swell from the size of a human hair to the size of a pea—they will become inflamed and will burn like crazy causing you much difficulty for walking; then, they may deflate giving you some reprieve. Neuromas are also associated with the feeling of numbness and socks wrapped around your toes and can be very debilitating. The treatment for these problems begins with cortisone injections that are supposed to cause the swelling to go down and control it for a time. The cortisone shot burns like crazy, and for me, either it didn't help at all or its effects lasted only a few days. (Note: The evolution of a neuroma is this: it will start by acting like a sock wrapped around your toe(s); then, inside your forefoot, it will become swollen and start burning and will become so painful that you'll swear you were lit on fire.) But like I said, it will for whatever reason remiss for a time; then eventually the fire will start, and it will not go out—now you really are in trouble; the pain is awful, and surgery(ies) in my case; removing the offender was the fix! However, the scar tissue left behind from the surgery may cause additional problems in the way of more neuromas, which for me would become another long-term challenge.

One night, several years after my lightning strike, I was again climbing the Great White Icicle in Little Cottonwood Canyon with my trusty Quasar ice tools. As I was working my way up to the top of the bulge—which is the hulking seventy-foot-high mass of ice—two-thirds the way up the route, without warning, I experience a full-blown neuroma attack in my left forefoot. The burning sensation was brutal and was not something that I needed while soloing steep ice several hundred feet off the deck. The pain was bad enough to break my concentration, which could cause me to fall and maybe be killed; in any case, I would surely be very badly damaged. I had to get control of my situation, so I banged in my ice tools a little deeper and tried flexing and arching my toes in an attempt to relax the problem; the pain worsened—

damn it, damn it, I yelled! My left foot now felt as if it were being clamped in a blazing hot-pipe vise!

I worked on getting my intensified breathing back to a steady rhythm then focused on getting my front points placed well. With my headlamp streaming light from my ice picks to my feet then below into the darkness, I started down climbing the bulge. I wiggled toes and yelled back at the pain and remained focused on good placements; it was difficult to kick my left crampon into the ice with that foot in so much pain. My left foot was a friend in trouble, and it was a vicious callous enemy wanting to see my end by making me fall, so I went into "battle and survival" mode simultaneously. I made sure all placements were dead solid and worked my crampons the best I could, executing precise crampon work in one theater while pain dominated the stage in the other; often they were in the same room! I thought to myself, watch one movie at a time, just climb, make it to reachable goals but keep moving. Getting to the base of the bulge was first on the agenda.

Long ago I had ingrained a mental stabilizer: when I was getting into a problem on a climb, I would look down then up to where I needed to go—to me, down was where I would go in the event of a fall, which is no good—up was where safety loomed. This process always seems to calm me down and help me become stable and deliberate again; tonight, I would find myself looking down to measured goals as opposed to up! The area at the base of the bulge is a lower angle and forms an amphitheater that is usually full of snow and represented a much safer place to be, on fire or not! Arriving at the amphitheater, I felt some relief 'cause dying tonight would suck, but the pain was still full-on, and I could sense a little panic knowing I still had several hundred feet of technical terrain to negotiate; I continued downward.

My next goal was to pick my way down the narrow and less steep third pitch (second ice pitch) to the next snowfield a hundred-plus feet away. Down climbing is a tedious and slow precise process requiring strict vigilance; I couldn't let my situation force a mistake. I did a fair amount of swearing and threatening to help keep my battle against the pain and my focus in check. Hearing and seeing water flowing under the ice was not a calming thing tonight either! By now, my left foot felt disabled, and the pain was taking my breath away; if I were in this painful situation in comfy street shoes, I would be on the ground with my shoe off trying to massage away the pain—no joke, but now, literally right now, I just wanted to cut my foot off. Finally, I had made it to the snowfield and felt some stress fade away. Still moving cautiously yet purposely, I used my ice tools to help balance me down to the final hurdle another hundred feet away. Arriving at the top of the first pitch, I turned to face the ice then carefully made my way over the lip and worked my way down this thirty-five-foot section of thin ice. Once on the deck, I dropped to the ground, full of anxiety, trying to rip my boot from my foot without undoing any laces from the inner or outer boot—the pain was a crescendo of stabbing and burning sensations. I swore, growling obscenities; I got up and went as fast as I could for my truck several hundred yards away downhill through the snowy forest! On the way down, my crampons helped bite into the slippery surface but also caused more pain by catching on tree roots and anything else their pointy teeth could sink into. Finally, I arrived at my truck, I hit the ground again, removed my crampon, and this time undoing the laces on the outer and inner boot, pulling them from my dying foot!

I spent a long while massaging the metatarsal area of my foot using a technique taught to me by my orthopedic surgeon, Greg A. I finally got a little relief and felt very lucky that I didn't screw up while high up on the ice!

Eventually, I learned that my snug-fitting footwear was adding to my problems and needed to address the issue if I was going to be able to manage this problem between operations.

For one thing, your feet swell as the day wears on, so I began experimenting with longer or wider climbing and backpacking boots to help alleviate pressure that may aggravate my existing neuromas, which by this point in time, I'd had three removed from my left foot alone. The operation involves a "four-week or more" hiatus from walking, and presently, I have been diagnosed with two more in my left foot and one in my right foot. These days I am extra careful with the shoe/boot/sock/insole combinations I use; if you pay attention, you can really stave off the problem and maybe completely eliminate the need for surgery. Eventually an opportunity came along to work with several footwear companies, which led to working with many sock and insole manufacturers developing products that would protect a person's feet, ankles, legs, and hips from the effects of the road and trail. I figured that if we could create products that helped my situation, it was sure to make other people more comfortable too! Honestly though, the manufacturers have most of the problems addressed, I just helped them tweak things a little and also helped expose the reasons for the proper fitting and marrying of boot, sock, and insole to the feet, or what I call the synergetic approach.

On the flip side of things, this was one of the many, many, positive things that came my way for some crazy reason because of the lightning strike—maybe this time, it was one of the doors that opened when others were closed, and I chose to enter it. As I have said probably too many times, like my feet, my neck was still on the forefront of my challenges and still hadn't been adequately dealt with, but I was also still dealing with anger and had been ever since I was alive enough to know the difference. Looking back just a few short months, I was mad at everything, including God and his buddies for hitting me with the lightning when my back was turned. I was mad at Mark for rescuing me—I hated him for it. I hated the medical staff for keeping me alive, I hated my friends for being friends, and I was sore at the psychologists for trying to figure me out, and I hated me! At first, it seemed that everyone had to help me do something; this was so embarrassing especially when it came time to go to the bathroom, it was humiliating. I wondered if the psychology department had a file and a smile for this. The bottom line is, back then I was different, and for a while looked different and couldn't think or reason, but somehow I gained an understanding about others in my shoes, so to speak! About a year after I left the burn unit prison, my dad, who travelled the western United States selling industrial chemicals, called me. Apparently, while visiting one of his clients in Washington State, he noticed their secretary had scarring and disfigurement on her face; he commented, "You haven't been around the last couple of times I have been in town." She told him that a few months ago, while riding her horse, she had been struck in the head by lightning and that she and her horse had both been killed, but she had gotten lucky and was revived. She then went on saying that she had lived in a burn unit and undergone skin grafting that left one side of her face something less than pretty to look at. She told my dad that she wasn't coping with the anger very well and was mistreating everyone she came into contact with including her husband and kids, and for no good reason. My dad told her a little about me and asked if it would be okay for me to give her a call sometime. When I talked with her, I was a little nervous; I wasn't sure what to expect except that we had been through a similar trauma. I shared my experience with this young lady then listened to hers; she cried a lot wondering what was wrong with her and why she was so mad at everything. It hit home so hard that I cried along with her, encouraging her and letting her know that it was just a phase that many burn victims go through and that with time, a whole lot of love, patience, and faith, things would get better.

She asked me searchingly, "Do you really think so?"

I answered, "First, you are alive, second, yes! You have to get through this for your family and your friends and for me. I need to see you make it through this ordeal!"

"Really?" she begged.

"Yes," I answered, finishing with, "You have to be strong for your family and the burn unit you lived in—everyone there needs to see you make it."

When we finished talking, I sensed that her well-being had improved; she really just needed to talk with someone who understood and would let her bare out her deepest feelings with no judgment. She needed to hear that she would be okay, and hearing that she was needed helped too! I am amazed at the power of words. Listening to her struggle put me right back in the front row to again view where I had recently been; in some ways, I was still there and I was still mad—I had simply moved on to a new set of things to be mad about. Amazingly my burns look great, and I will go through life with very unnoticeable scars even though I have a big one that occupies the area just behind my face. I too will have to face what she must!

As you may recall, I mentioned Jack recommending that I read *Psycho-Cybernetics*. In that book somewhere, Dr. Maltz comments that he can do wonders with plastic surgery and make some of the worst damage seem new again, but what he can't fix is how the victim, no matter how good they may now look, sees themselves—therein lies the danger and the problem. I really hoped that this young lady would be able to accept herself again and move through life reasonably well. My own anger and the fact that I saw myself as being different would follow me like a ghoul and make life and survival difficult for decades to come! Eventually I would become surprised that I survived this one single problem. I thought back to the comment I made to my wife: that if I could get my various sources of pain down to a manageable number, I think I could survive this thing. I didn't realize just how powerful some of the "manageable number" items could be and that several of them were, on their own, life threatening in their own right! Of course, this manageable number of items had to involve the obvious (i.e., neuromas, my ear issues, my spine and neck, joint issues, and finally my anger, which would tail me forever); also lingering in the background, there were the other questions from the so-called Black List: brain, organ, and nerve functions; would they survive, would I? What a haunting thing to have to worry about. I chose to focus on goals that I wanted to reach despite being mad and frustrated about so many things; I would work the issues as they presented themselves, and like it or not, I just had to keep learning and accepting the art of adaptation.

Two decades later, while out hiking, I am ever aware of my limitations and am able to work with them while having fun and enjoying nature; I am able to reach reasonable goals, and it is nice to have goals to reach and to be able to see them through. I am so damned lucky, but I wonder about how the others have managed their issues—the girl from Washington, where is she now, and what about my burn unit comrades, did anybody really survive?

It was pretty clear from early on that my neck was in for the long haul; it had even made a place for itself on the Black List, which clearly spoke of other dilemmas that I may have to confront, including tendons and muscles that may be more easily damaged due to the effects of the electricity that ran through my body. I guess even despite my stubborn, independent, and stupid nature that when you are so outnumbered (a fair match in my original mind), you just have to be very adaptive and have dreams and goals and see what life will allow you to still have, at least for the time being. This is reality even though it gripes me to say such a thing since it sounds as if I am giving in. But one does have to be smart and pay attention to the signs along the way if one is to last very long; every day—in the end—is a gift!

This neck business—my neck, that is, and its issues were on the surface or near the surface from the beginning and would become the most obvious significant long-term injury of the bunch that I would have to deal with. This one thing would cause me to change my life and way of doing things and what I would and wouldn't experience forevermore, lest I try going against the ordinance of higher powers and die!

Reflecting back to when I got zapped by the lightning—my injuries had brought on things like the following:

a. Weak muscles etc., making it hard to hold my head up.
b. I had and was damaging nerves in my spine and neck every day as I worked and played.
c. It was hard to work and to climb as I experienced numbness in my head and in my hands, intense headaches, and momentary loss of eyesight especially during times of head and neck articulation.
d. Stress and balance issues.
e. Emotional distress and questioning of self in technical or highly physicals situations.

This was difficult to accept because of how disciplined and sure of myself I had been before. And now, those around me wondered about me: was I okay to be in the real world, even the doctors wondered. The stink of it was, is that I knew all of this, and I questioned me too! All of this worked pretty hard against my growing determination to get back to the old me, such a war inside and out! Everything including my own mind and body was trying to overcome me.

Thinking back to what my friend Ted, the recruiter, said late one night in the burn unit about my new job—I was to go out and tell my story to others, to reach out and help them. Me, help others survive their events? I wondered how much good I had done by talking to the girl in Washington; if she could only hear me right now, it may ruin her for good! I found out, however, that no matter what my neck or feet or burns or what have you was doing to me, that when I did reach out to help an individual or a group with their unique struggle, that I received help too! I always felt better inside and out and realized each time that I have to keep working through my problems so I can help others through theirs. Inside I knew through faith that I had helped everyone that I talked with or had recovered for; I knew that through faith, the right seeds had been planted and that each need would be filled at the appropriate time.

The neck issues I listed were analyzed by my current orthopedic specialists for about a decade and a half. I always seemed to wind up getting some meds in the form of pain pills and muscle relaxers and working through many rounds of physical therapy. I sensed that I was decaying physically anyway and maintained a foothold on my health by remaining very active. I believed, and I still do, that it was the only thing keeping me from the uncertainties that were waiting just around the corner. I firmly believed that if I gave in and slowed down, then harder times were in store for me. I was afraid of what was happening to me, for I knew that one day I might just break down and that would be it.

Sometimes I wished I had gotten a little more mental therapy; inside I felt so alone and knew that if I complained, I might get removed from my life. So I had to press on like a good soldier and fight the menace—even inside myself. There were days when I was so scared to do my job, particularly preventative maintenance on energized gear (i.e., electrical panels, motor controllers, frequency drives, and transformers). Back in them days, we didn't have arc flash suits to protect us from a fiery arc flash—you just didn't make that mistake! If you did, you paid by absorbing some level of arc flash and associated burns, or maybe you got lucky with the scare of a lifetime; none of this is good for the soul. Fear—real or imagined—is real and is very hard to describe its debilitating effects! I just know that being afraid paralyzes me and makes me feel like a loser, and it hurts, and it's hard to admit, so you hide away in that little hollow place inside yourself and keep quiet or else. I think some of these experiences hardened me a little, building up a brick wall that became my game face and who my surface person was; I can't say that I always agreed with this perceived toughness as it caused me to take risks that I shouldn't, and with every success, it became more okay to do so—when in reality, it wasn't okay. Our leaders were just fine as fine could be with the risk taking as it often saved time and money and made them look good! I guess I decided to fight many of my fears by facing them alone when I could, which was most of the time. The only thing I can say in my defense was that I was very considerate of the objective dangers and was very careful during a risky process, but no matter how skilled I became, agreeing to such risks is really very selfish. For many, the big picture never comes into view; it's always about that small moment in time. There is good fear that should make you respectful, but it's the bad fear that I have seen nearly or completely destroy somebody.

A few years back, I was again soloing the Great White Icicle. I was in good form, feeling great, and in total control of all the things that I could be in control of and ultimately really having a good time. After I topped out, I made my way up the stream that forms this beautiful cascade; after a few yards, I came to the point where you turn west to get to the Descent Gully. As I began crossing a very exposed ridge lying between me and my gully, there sitting in the snow clinging to a bush was a teenage girl, a climber I supposed, because she was wearing crampons and a harness. She was crying, and I could see she was mentally fractured and scared to death to make a single move! I don't blame her, this girl was clearly not in the right frame of mind nor was she skilled enough to be where she was or to make the next correct decision. As it was, she was facing a death fall of hundreds of feet, and in her spot, it is very omnipresent! I asked her what her name was, sobbing, she said, "Kim!" So I said, "Kim, I'm going to help you get to safety." She nodded saying, "Okay." I then tied her in and put her on belay, helping her over the rough ground to a flat area where she could get a hold of herself. I asked her how she had wound up over there on that ridge. Kim told me that she and her boyfriend (so-called) agreed to meet at the Great White Icicle to climb that day; then, he (the hero) would lead her up her first ice climb. Apparently she and her so-called boyfriend had climbed the icicle, and she had not been moving fast enough for him! After they topped out, they untied (not so bright with a beginner in tow), put the gear away, then headed west for the Descent Gully. He left her alone on this ridge to make her way down to the valley floor.

If nobody had come along, she would probably have frozen to death, fell, who knows which, but something bad would have become of her. I agreed to get her down safely to her car if she promised to leave this jerk then tell her dad and his dad and all his friends what this little creep did to her! She promised she would. I also told her not to go ice climbing again until she had gotten with the local guides for some instruction and guidance! As we made it to the parking lot, she mused that her boyfriend was gone; too bad 'cause I had a few pieces of advice for this little creep. I gave my young friend a hug, wished her the best, and we said our good-byes. I figured we had seen the last of Kim in this land of cold, steep rock, ice, and snow, but wondered how this experience may affect the rest of her life.

With a slight bit of a twist in contrast, here is another example of how fear works its worms.

While working at the Air National Guard as one of the civilian electricians, many of the employees began noticing that our secretary was showing up to work with black eyes, bruises, split lips, and just generally beat-up looking and becoming withdrawn.

The look on her face was usually describing fear, question, and dread. She did her best with makeup to hide the damage, but the cover did little to hide the swelling and quietness. I and a buddy, Tim, asked her what was going on; at first she avoided the conversation, always being too busy to talk, but we pressed the issue because she was our friend and she was being hurt. Finally, needing to talk about it, I think, she opened up!

Her husband had been beating her physically and mentally for over twenty years; he had told her over and over that she was worthless and was stupid and was nothing but a loser and could never make it without him, that she couldn't spend five cents without his permission and she wasn't allowed to think either except by way of permission, he had made her afraid of him! This predatory animal/psycho made sure of this with his feet, fists, and his mouth. Their three kids (all boys) were tortured together and individually along with the surrogate punching bag (their mother) often being abused right in front of them. She told Tim and me that one day she woke up and decided that it was over, that she was going to escape from this "warden and his prison and the nightmare"—she had to! So she confronted the "warden" and told him what she planned to do! He did what any dictator would do; he amped up the punishment, which was what we had been noticing lately! Tim and I were furious, we offered to confront her husband, but she declined our offer insisting that she would handle it herself. This gal eventually left her husband and divorced him, but he still, despite court orders (paper walls), threatened and beat this poor woman. All the court orders did for her was get her some more black eyes and split lips even after she called the police. Remember this, the cops are always too far away when the intruder is on your doorstep. You can be beat silly and killed long before help ever arrives; this is a fact, and she got to realize it all too often. Trying to prove he was even there seemed to be impossible, becoming hearsay. Her pain and worries were not hearsay though.

Eventually this tormentor went away with the wind, and this woman wound up working two full-time jobs, bought herself a car, then a house, and was able to make a life for herself and her kids. Initially, when she began creating this new life, she looked terrible because she was hardly getting any sleep. Often she would come to us technicians for advice on how to do a home repair or fix something on her car etc. We would offer to come over to her place and help, but she would politely decline, saying she had to do it herself! This woman had suffered greatly to gain her independence and needed to maintain it at least for now. This gal finally did find and marry a really great guy who was one of our guard members! Rod died of cancer a few years after they wed; she had taken really good care of him all along the way, and he gave her the greatest respect he could by just loving her, even if for a short while! Sadly, this woman died of cancer not long after her new husband passed. She was loved by a great many people and was known as a giver, not a taker! Looking back, I was amazed and humbled that she was able to eventually, in the face of intense fear and beatings, survive all of it, and then do well for herself and her loved ones. This woman survived because she had a reason to live, it was for her children and her grandchildren and for herself, and so she did.

What blows me away is how commonplace the negative and brutal part of this woman's life actually is! You can't take ten steps in any direction without hearing similar stories of abuse. Why is this so, where do we learn these manners from? And what about all the other things that occur, like rape, murder, etc., and I'm just skimming the top; and this is the USA, not some freakin' place where they think it's still 1500 BC.

Sorry I am saying this again, but in my case, like her kids,

a. I was born, and then I got to witness the beatings of my mom and siblings with all the screaming and yelling and pain.

b. I got beat up too; everybody gets a turn…is it out of fairness…
c. my siblings and I hid in neighbors' houses to avoid the bigger beatings of bad days. I was only five years old at this point. By the time I was seven, I had suffered broken noses, a concussion, and several broken vertebrae, and by accident have nearly drowned in a local lake.

Most of these were caused by a mean parent.

By now, screaming and yelling and arguing make me sick to my stomach, and I want to hide, I want to find the dark corner of a closet where nobody can find me; then I want to hide even deeper in that hollow place that is like a deep cave far into my mind. Looking way back, I could see why I had become withdrawn, and school was becoming difficult, and I found myself struggling in important topics like Math and English; I knew I was failing, and I felt stupid. I felt like a loser because of having to live needing welfare to eat and Sub-for-Santa just to look seminormal within the community. In the sixties, not having a dad set you apart, at least where I come from; you were different and treated different, and it hurt! Whether you had some money or no money, it could keep you on the sidelines. So I barely made it through school with little chance of succeeding in life, I thought. This was a dangerous seed that had been planted deep in my mind. So what do I do? I choose to move back in with my dad so I can experience his version of life again. I guess I had better put some more glue on those hurtful feelings lest they fall apart.

Then sixteen years later, I get hit by lightning and get all torn up, but I am supposed to try, I am supposed to see the light at the end of the tunnel, and on top of all the other crap that came along with the last sixteen years—why, for what?

In the end, it is so easy to see only the negative side of any situation, and it is easier to just give up and not consider the toll your actions will have on yourself and others.

Our secretary knew—like my mom knew—they and their children would be hurt over and over, their lives were at stake! Eventually, both women, despite their great and many fears, found the courage to leave their tormentors despite threats of more harm. These two women and many others like the burn unit victims who were rooting me on, weathered the storm. They found reasons to survive, to live life, no matter what was handed to them as status quo. And me, I think I finally realized that I focused way too much on all the bad things that had come my way. I let it build impenetrable walls between me and the good things in life, like love and people and life itself. I added to the barrier that was slowing or stopping my progress in the direction of getting better.

I could see that even though outwardly it appeared that I gave of myself in some material way—for example, I have shared my lightning story with hundreds of groups, including Boy Scouts, electricians, military folks, and just everyday people—I firmly believe that there is message contained therein that will help somebody somehow. I make sure I donate to charities—be it the Boy Scouts or homeless groups, and I have helped with many home projects and so on. I have also given rappelling classes and backpacking classes to Boy Scouts and military folks—somebody has gotta show them right from wrong! I have helped the military with many unique projects, utilizing my free time to research and get the answers needed to help see the plan carry forward and was always involved in helping the taxpayer get the most bang for their hard-earned buck. I've consistently made sure to give my wife roses often to remind and thank her for all she does for me and our family! Yet amazingly at the same time, I never let anybody get very close to me. I was always afraid of how my limitations would make me look to others, so I've stayed hidden to a very large degree. People may notice that all too often, I am the only one in the picture while I am out field-testing or playing a guitar. I am pretty embarrassed about myself and need to learn to accept me! As an illustration, I have conveniently found reasons to not participate in group functions or with friends or family. There were times when I would be made to work during company Christmas parties, and this seemed to cement my isolation needs even more. I have turned down opportunity after opportunity to go on road/climbing trips. What was wrong, was I afraid I would have a good time? It's good to break away even if it's for a weekend to be with friends. However, I have been on a few group road trips, and the sense of adventure felt great! The newness of it all adds to the Great Spirit that climbing or backpacking exudes. Keith, Fred, Tony, and Bob—all of them very patient and faithful partners; I pulled the plug on many of our adventures, which remained in the dream state that hatched them. That time of opportunity went away with the wind and the dust and the smells of the changing season, and for that, I am truly sorry. I wanted, I guess, the security of being alone! Sadly, all my friends wanted was nothing more than a little company.

I am surprised at their confidence, that they would still want to climb with me after the saga of the lightning strike! It was so difficult to overcome the fear of question; it really was a question of, did I really still possess the nerve and reason to come back to just this one activity? Thank God for patient partners because I did possess the nerve, but in my self-restricted mind, only a little. I will say this again: so much good was happening to me and all around me; you gotta work through your problems and let others help! The process may just help you, and maybe, if you are lucky, somebody else will be better off for it too.

Embrace life and others. There is always someone whom you would not want to trade places with—sure, I've had it hard, but my issues compare not at all to the damage of a war zone or to my friend Dennis who lost a dad one Christmas, then his mother the next, and the next a brother! My second oldest brother, Jesse, lost his mom, then his wife, daughter, and best friend over the course of five years; my burns hurt, but not that bad! You know, realizing that you have a problem is half the battle to fixing it; the question is, how bad do you want to? This question again makes me think back to Steve Anspah (the recruiter from Reno) telling me that I had to find one reason to live—not two, just one! For some people, life seems easy; they simply decide, and everything goes the way they choose. For others, there seems to be a longer process of just getting to the point where you want to live, let alone deciding to fight for your life. All these differences need to be taken into account so people can be understood better and get the help they need to work through whatever is holding them back.

In the past, I found it easier to sit back and do nothing as it seems safer and, to a degree, more comfortable, even if it meant never walking again, like in the early days of my recovery. Because taking the next step often means stepping toward the unknown—will I get better, or will I get worse? What about complications? Will I die on some operating table? Remember the close call I had while having my left foot grafted! A simple allergic reaction to the anesthesia brought me very close to the end, leaving me with a deep-seated fear of going under the knife, which from then on would cause me to panic as I began the elementary count backward from one hundred. I believed that when I finally went to sleep, I would never wake up again! I realized that something deep within me wanted to stick around to see how the movie goes as opposed to walking out halfway through! Two decades later, after looking at all the self-analysis I engaged in (i.e., back and forth; left turn, right turn; give up or keep going; live or die?), I really must want to live 'cause here I am. I know I know, I have repeated this way too much, but good things were going on all around me. The universe just seems to be about balance, and so it is with life!

Despite the many weakness I had accumulated from the accident, I had gained—or was gaining—many strengths too, but these strengths could or would not rear their heads for a while. Wading through the rubble of the accident trying to make sense of all my various feelings, I knew that I projected many of my feelings onto various people's shoulders, and it made their lives more difficult at times and at other times perhaps more positive. Life for me had become physically and mentally harder. My mad stage (born from mostly fear, I think) has endured through the years to one level or another. Fear of failing and/or change has kept me a bit sore; for better or worse, it has kept me in the war/survival mode all this time! However, at this late date, my inclination is to win the war at all costs by staying as healthy and strong in mind, body, and faith as I possibly can, while still being fairly reasonable about the whole affair. I firmly believe that by doing this, I will win the war against anything life brings my way, I must endure. Lastly, it is all that I can do for myself, it is all I have left and represents my final options, which means that if I am to have any quality of life against the degradation that I was told would happen to me because of my accident (which doesn't factor in natural aging), I must maintain this mind-set. Well, crud, now I gotta get old too, well, I ain't doin' it, I am staying the age that I am right now, and that's all there is to it! This process always helped me reach goals before, so it should serve me now as well! A long while ago, my grandpa Wally told me something; he said that it hurts to get old even if you don't have injuries to contend with, so whatever you do, stay in good physical and mental shape, and guard your faith fiercely. I have never forgotten this advice and now am on the pathway of realizing just how true the advice is from that old grandpa of mine. Keep up the faith.

13

Trade Shows and New Friends

Throughout the early to mid-1990s, my creative mind began waking up; as a result, I began writing lyrics and music again. I also began getting ideas for things that could be useful to others in the outdoors! A National Guard buddy of mine, Rick, and I collaborated on an idea (I needed his money) acquiring a design patent for an invention known as the Rifle Backpack! Acquiring a patent is no small feat and can be quite expensive, involving patent attorneys and so on.

The rifle pack was designed to safely carry a hunting rifle on your back and out of the way so your hands could remain free to manage your downed game. To our surprise, the patent search clearly displayed in black and white the fact that even as far back as the late 1800s, patents had been issued for ideas quite similar to ours. While we were trying to figure out how to market our idea, we became aware that several backpack makers were already marketing a similar version of our idea but used for a different purpose; nonetheless, it was time to confront these skilled and savvy businesspeople. The lightning bolt had done something for me! Even though I typically choose to hide in the shadows avoiding confrontation, it had made me tougher, and these days, I am not opposed to standing my ground if I'm forced to. As I saw it, I had been dead and burned up already, so what are these companies gonna do to me that could be worse? We decided to send Wayne Gregory of Gregory Mountain Products (one of the apparent offenders) a formal letter concerning the issue. Several days after the letter had been mailed, my wife called me at work. She told me, "You've done it now. That Gregory guy called and wants you to get in touch with him ASAP." Whew! I thought, *That's great!* Now I gotta debate this issue with a savvy business guy, and I'm probably toast, I couldn't even get mad. Ultimately, I sensed that Mr. Gregory probably wouldn't be too mean to me and I just might survive our pending drama. So I gathered my wits and returned Wayne's call! We quickly dispensed with formalities, the attorneys on both sides agreeing there wasn't any infraction given the way the idea was being used by Wayne's company or any of the other companies for that matter. However, since I am the creator, I still disagree with the decision; I'm ornery I guess! Through all of this, Wayne and I became fast friends, as we were, no matter what, kindred spirits; we both love backpacking and both are incessant gearheads; we love fishing, good food, and we both are people persons. (Note: Isn't it amazing that I have been saying all along that I hide away from people?) In as much as the lightning made me ashamed of myself, meeting Wayne would cause me to meet thousands of people with whom I would frequently correspond and work with and hang out with. Hmmm.

While Wayne and I were getting to know each other better, I gave him a healthy dose of my keen interest in backpacking and its required gear. After I got done running my mouth, Wayne asked me if I would be interested in becoming a "wear tester" for him; he said he had testers all over the world and needed someone in Utah. I said, "Sure." Wayne then told me that he was gonna send me a three-hundred-dollar backpack for me to test.

Emotionless I guess, I said okay!

Wayne hesitated a little then replied, "You don't sound very excited…"

"Oh, I get it," I said, "you think that I should be doing backflips up and down my kitchen here 'cause I am going to get a three-hundred-dollar backpack!" I continued, "The truth is, Wayne, I am excited, but I presently own a Gregory Snow Creek, which is one of your top models, and I have owned a dozen other top-notch packs. My job with you will be to work and help create even better backpacks, not to test equipment for the purpose of acquiring more equipment."

I think this surprised Wayne; he quickly realized my honesty and diligence was for real, and we went to work together perfecting his awesome backpacks. Wayne invited me to the next Outdoor Retailer Show, which is where manufacturers and retailers get together to see what is new for the next season and place orders; the show is held in Salt Lake City twice a year, once in January and then again in August! I would show up several days in advance and help the Gregory crew set up their booth, which involved constructing displays that held all his current backpacks and duffels. From the lighting girders, we hung giant silk screens showing famous mountaineers using his products while climbing Mount Everest; we also assembled lighting arrays to highlight all the products and the business floor. Wayne, through the show medium, introduced me to many diverse and interesting people within the outdoor industry; Wayne knows and is known by everybody! One time Wayne commented to me: it's not *what* you know (which is very important) but it is *who* you know. Get it? Such is life. Connections are a key to survivability. My innate independence wrestled with this idea lots, and still does, but I can't argue with its truth. Wayne also gave me my business name:

Lightning Layne's Professional Field Testing; then he sent me out to meet people around the show and see what I could do. I would introduce myself and then tell them about my background in climbing and backpacking, which I had been involved in for about twenty-five years collectively! But it was usually the lightning strike story that got somebody's attention, breaking the ice so to speak; eventually they would wind up agreeing to have me test their products. Most people were blown away with the story, and even more so when I would show them my sock, which was burned in two, my destroyed hat, and my torn-up pants! But the real attention getter was my Asolo Trail 2 S boot, which had been blown off my foot, sporting its two-by-ten-inch lightning exit hole where the top of my left foot went through. This boot really raised eyebrows, becoming my real calling card. Many people saw me as a miracle; I was embarrassed by the idea though and usually referred to myself as being supremely lucky. Most of these folks wanted to hear the whole story, which I have always been happy to tell; somebody once told me that it was my new job, remember? Often the person(s) I was relating the story to would grab other associates to come and listen too. On one occasion, I went to Columbia Sportswear booth to meet their product manager, which for any company is who I wound up working with. As it so happens, I was wearing their stalker hiking hat when I got struck by the lightning bolt, which blew the hat into four pieces. The first person I met at the booth was Ma (Gert) Boyle, the towering gruff-looking woman in the ads who really is only about five feet tall! She asked if she could help me, so in a serious tone, I asked her a trick question. Showing her the pieces of my hat, I asked what was wrong with their R&D department; couldn't they make a hat that could handle a little lightning? She gasped then asked what had happened, so I introduced myself and gave her a sentence or so version of my story! At that point, she told everyone in her booth, including clients and employees, to hold on for a minute and told them to listen to me, that I had something to say that they needed to hear! I then gave a more detailed version of the story. I got to know and work for many folks in the outdoor industry by this method, and many have become lifelong friends.

As it so happens, I was wearing Fox River backpacking socks when I got blasted, and the people at the Fox River booth just took me in like a poor stray that had been left on a stranger's porch! Candy and Jenni had me tell my story to their crew and put me to work helping them develop the great hiking and winter sports socks they are known for! When I made my way over to the Asolo boot company, they were somewhat less than friendly! When I showed them the boot asking if I could become part of their field test team, they made sure that I was aware that they had European experts that took care of these details. I then asked if I might not get a replacement (even used) for my left boot since the right boot was still serviceable, they flatly said no! They also insisted that no lightning bolt had done the damage to my boot and suggested that I let my dog (which I didn't have) chew the daylights out of this boot, and then they shuffled me on my way! Obviously my story and its examples were not gonna win over everyone. However, life is about change, and things evolve; glaciers inch their way down valleys, rivers dry up, and boot company managers move on. Asolo boots at this time were getting some pretty bad reviews and needed some help! Amazingly, a new management team came into place for this company, and I was given the opportunity to work with them helping to develop ice-climbing and hiking boots! The Zenatta brothers, Marco and Luca, introduced me to their product managers Stu and Jason, and we went to work making things better. Of course my interest would be widely directed to the overall performance of the footwear, but I also needed to cater to my left-foot issues so I could stay in the game of ice, rock, and mountain trails a bit longer! The challenge would be to create footwear that could help me be more comfortable; in turn, this most assuredly could help many others enjoy the outdoors too! It occurred to me that this venture that my friend Wayne had exposed me to was a bit of a dream come true; it also helped me come out of my shell a bit as well.

I love the outdoors and how it makes me feel, but I am also a real gearhead; I love the equipment—there ain't anything that I have one of when it comes to outdoorsy stuff. I would like to take Jesus for a hike and show him some of the cool stuff we have compared to putting up with some stinking camel or a nasty old billy goat! Anybody from a hundred years ago and back would be blown away—take them out for a subway sandwich and watch their eyes get big! I can tell you that my backpacking sleeping and cooking gear would make a sheikh's jaw drop—finishing him off with an F18 hornet screeching by one hundred feet off the deck. Isaac Newton would have a heart attack if he saw the telescopes available these days—game over, we got it all, life is good, and I gotta stay in it.

Over the years, I've had the opportunity to develop short- and long-term relationships with companies that produced equipment that could help me (especially with my new limitations) reach my goals, enjoy myself, and to develop me as my new self too. I am pretty much still the same guy I've always been, but I was battling this new set of circumstances that were piled right on top of the junk (as I saw it) from my childhood. Working with these folks was a way to help everyone, especially those in my shoes, have a better way of life.

Here is a good rule to follow: Do you want to fix your problem(s), help somebody else with theirs. Consequently, I worked with several boot companies and virtually every insole and sock company that came along! This work also led me to several other backpack brands (i.e., Lowe Alpine and Dueter). I have even tested watches; amazingly, they told time, testing me and my timeliness. As an independent field tester, being objective is a major prerequisite; being diligent to protect the interests of the competing companies I worked for was absolutely paramount, and I know that I trod on hallowed ground (with permission and great trust) where very few, if any, had ever done so before!

Mountain Safety Research, better known as MSR, product manager Georgia Bakke was quite direct about the subject of trust!

"Why should we trust you with any of our products?" she queried. "You work for so many conflicting interests."

I thought about her question for one nanosecond, then replied, "For one, you could sue me, and two, you may compete viciously with some of these other companies, but you are all in the same business and are mountainfolk! If I were to do something to hurt your business, you folks would get together about it, and I would be completely finished as a field tester! Georgia, it all comes down to personal integrity."

We made a handshake over the phone, and she trusted me to work on some of their products for which I was very thankful. Now bear in mind, when I say "work for someone," this implies that I was on the payroll—yes, I was part of a professional athlete program, but in the nonpay status; I never made a dime testing gear. In the beginning, Wayne told me that because of my unique place in the market that it needed to be this way! So I accepted the rule and did the work for the love of it; it was cool too, because I could as I should be—always unbiased. I couldn't be fired, but could be asked to just beat it! That never happened however even though I know there were a few people who were less than happy with the truth that came from my reviews. It was my job to be honest; it is the only thing that really makes a difference. I also had the opportunity to work for tent makers, sleeping pads makers, and stove makers; I even worked with heat pack / body-warmer makers (you know those little disposable "shake 'em and they heat up" pads). I met so many phenomenal people all along the way! I have one regret however that I must talk about, especially since I was just dealing out the ethics sermon with a big ladle!

I violated a rule one time, which I will call professional honesty! I was testing a hybrid tent for Sierra Designs; it was designed to be a lightweight assault tent meant for the extreme environments of the Himalayas. I had taken it into the Uinta Mountains in northeastern Utah in late fall to see how it faired in that climate; well, this tent had some issues! The project manager Fred had implored me to Seam Seal that tent before I use it. Interestingly most tents come Seam Sealed from the factory, but not this one; I refused to seal it because what this meant was there was something up with the fabric or the seams or even the thread, and I had to find out what was wrong. My friend and Guard buddy Robert Cole and I spent a miserable drippy wet time in that tent during an eleven-hour downpour; we were soaked. The material passed water very easily especially if you touched it; I believe every needle hole let water through too.

This tent also caught wind like a sail; one model blew away while being tested in high winds of 90-plus MPH on New Hampshire's Mount Washington observatory deck! Fred was watching the tent struggle through a window from inside the observatory and was telling me on his phone, "I wish you were here to see this," and then he exclaimed, "it's gone!"

Startled, I asked, "Gone where?"

Fred simply repeated, "It's gone, but its stake loops are still out there flapping around!"

This tent's frame structure couldn't handle much snow either. I managed to break an Easton aluminum 7000 series pole with just six inches of snow loading! I was about to let Fred in on the facts of my test (my job) when one of his assistants told me that this tent was his dream, that he was very excited to see this tent succeed—for that to happen would require new materials and a different design. I had come to like Fred a lot; he had been so genuine and ready to let me work with him on various projects; how could I kill his pet project, tell him that I broke the pole of this $700 mountaineering tent in a local canyon at moderate elevations with minimal snow, that every one of my climbing friends who reviewed the tent said it would never handle wind or rain very well? But Fred knew that much! I really thought the project should be reengineered or scrubbed, so I softened the facts a little, saying it had a place in the mountaineering world (it could with some changes), but as it was, you would have to dig it in deep to protect it from intense wind; snow loading would also be an issue requiring the party to do extra work to protect the shelter. The project died on the vine on its own in a very short time before it went to market. I failed Fred; I should have told him sooner to scrub the project or look at a redesign. I truly regret not being more forward and for letting emotions get in the way of me doing my job. I promised myself that I would never do that again, and despite the ringing in my ears from the lightning event, I could hear a different ringing!

It was the advice Wayne had given me at the very beginning of all of this, which was to be honest and always tell the truth! This doesn't mean that the input will always be accepted with happiness and joy, but it will always be the right thing to do (nothing like good old wisdom to whip ya into shape and humble you), and I did so with one of Wayne's designers going a bit too far! I had tested a pack for John that was intended to be a crossover type that would cater to many sports—it was truly ingenious, and my son Eric carries it to this day! I think I lost my objectivity with this project. I complained insidiously about the waist belt not handling an ice climber's load, which is quite heavy as you may carry several ropes, all your ice screws, and rock protection, ice tools, food, extra clothes, water, etc. In truth, this pack handled my fifty-pound load about as good as any other alpine-climbing pack, which typically has a scaled-back harness system as compared to standard backpacking packs. This pack worked well enough for its intended purpose, but like I said, I wasn't being objective or sympathetic to the purpose of the design. This sort of evolution by the way is how all the great gear we enjoy comes about in the first place. I'm not sure what I thought this small lightweight pack should do, but to a degree, I was wrong and had clouded my own creative view by being too critical and could have been more useful to John if I hadn't been so heady.

Note: Ski/climbing packs are designed to be somewhere between day packs and bigger, more robust backpacking packs. They simply are accessorized to carry skis, ice tools, and other alpine-related gear in a more utilitarian way and are typically built to endure a rough life. Quite often these packs even double (if the design permits) as partial bivy sacks covering one's legs and butt if one gets stuck and has to spend a night away from the tent or house. The haul weights these packs and their mules endure can get pretty heavy, but happily this is usually shared between partners (mules) before the gear is transferred to the activity itself. Around this time (the mid-2000s), I was to a degree restored physically from the effects of the lightning strike and had become a very technically capable climber and tester. I was also not being very open to the new kid on the block at Gregory Mountain Products design team. To date, John and his accomplice Mark have created some of the most technically advanced backpacks known to man—hats off! All of this reminds me of a story that I was already very well aware of but apparently hadn't learned from its lesson—again, or yet?

The Gore-Tex company knew what they were doing, especially with how they dealt with people! They believed that everyone has good ideas and were very adamant about recruiting ideas from everybody on their staff to help improve their products. Apparently one of the maintenance staff members was responsible for the Seam Sealing process that became standard on products containing Gore-Tex; this points out that everybody has something good to bring to the table!

I also got to do some interesting things in the thermal underwear part of the industry too, testing clothing for Terramar, Mountain Hardwear, Optimer, and Minus 33. During the early stages of testing for Terramar, I was teamed up indirectly with such visionaries as Jeff Lowe and Will Steger from the climbing and polar exploration worlds. Compared to them guys, I was very average, which might have been the planned contrast from a wide viewpoint perspective. Back in them days, synthetic underwear (particularly polypropylene) had one very nasty attribute! After you had worn it for a few hours, it stank to high heaven; your partners couldn't stand you and vice versa, and the idea of being in a tent together sucked; people down the trail could smell you coming; billy goats turned up their noses at you. I guess the only benefit was it helped keep you dry and warm, plus it probably offered a deterrent to grizzly bears and mosquitoes. Terramar was trying to do away with this terrible odor at all costs. My point of view to Russ, my contact at Terramar, is that odor is odor, whether it begot in the far north Mushin to the Pole or rest stepping and pressure breathing to the top of the world or on some local ice climb, cross-country ski trip, or backpacking trip! Of course, each one of these environments offers their own unique challenges for the fabric in the way it will manage moisture to include keeping you comfy and warm. So I think this idea is what helped me become part of the team, plus I wasn't on the payroll—I represented the everyday Joe! Will and Jeff, during part of their test, wore the same undies for many, many days; I don't remember how many, but it was a lot, and they claimed they did not have an odor problem! Really, says you, says some? So following their lead, I did what I always do; I went completely and officially overboard. I ice climbed, hiked, snowshoed, and cross-country skied in my Terramar undies and did not wash them for half a year—yuk baby, but I agree with Will and Jeff; the odor control works! Russ, the front man for Terramar, asked me if this was all true, I said yep, I've no reason to lie and nothing to gain by not stating the facts; he then invited me to say my piece to a large group of Textile folks associated with this product. Some were impressed while others' jaws went a slack and probably decided to not invite me to Thanksgiving dinner either. I laid it all on the line about my dirty laundry from the "regular guy" point of view, which represents a huge part of the market these folks would sell this clothing to—this product works.

It's worth noting that Boy Scout Troop 440 of West Valley City, Utah, became the first trainers for lightning awareness and safety (it was Wayne's Idea) and created a booth for this cause at a recent Scout-a-Rama. Because of these Scouts' efforts, the Terramar leaders Russ Pittman and Jeff Danzer supplied this entire troop with cold-weather base layers. Another company, UST or Ultimate Survival Technologies, supplied fire-starting tools and whistles to the troop as well. Incidentally, Wayne Gregory gave a backpack of their choice to any Scout in his local area that made it to Eagle Scout—pretty cool, huh! This demonstrates the type of folks that occupy this industry. Eventually I managed to become part of field test teams for the North Face, and Nike too, working on trail-running gear! I found some of these companies difficult to get into at first, but patience is a virtue, and by adding some persistence, I manage to work with many people who initially shooed me along.

One area of the industry that I put in a fair amount of energy was helping to create and test gloves specifically for ice climbing. With glove makers, this was a tough nut to crack because it was so specialized and there were already gloves out there that seemed to work pretty good, but to a skilled ice climber, this was a half truth. We needed something better!

Ice climbers need as much dexterity as they can get; warmth, waterproofness, and durability are also pluses on the desired list for the "man of the ice." Long ago, one of my ace climbing partners, Fred, told me to get up to the Black Diamond store in Salt Lake and buy a set or two of their Kevlar-palmed mountain gloves. They were the bomb! BD, or Black Diamond, had a basketful of them and was closing them out for forty buckaroos, which was nearly half price. I bought two pairs, one was hot violet and the other deep purple, and I noticed that the palm area where the Kevlar was seemed very indestructible. The fingers were very radically curved, actually matching the natural "at rest" curve of the fingers, which made it easier to grip an axe and to fiddle with bineers (carabineers) ice screws etc. The shell was some version of a proprietary waterproof nylon and contained a thick fleecy removable inner glove. We had to seal our own seams with Seam Grip; otherwise, the glove would leak and let water in to soak and freeze our hands up. We also installed idiot leashes on these very simple gloves and was now the proud owners of the "Cats Whiskers" of gloves, at least for the time being.

I used the purple gloves for ice climbing because it contrasted well with my mango-colored shell gear, purple Misty Mountain harness, and BD (Black Diamond) inferno-purple boots. Being so color coordinated like Fred, I didn't look homeless anymore. Fred used to say I may be a terrible climber, but I always look good—to me, he belonged in a climbing shop window on display! The BD glove wound up teaching us lots about an ice glove's fit and function. One very important issue was when it came time to rappel, even performing a very average slow methodical rap or rappel could cause (due to friction and thus heat) the rope to burn through the nylon or rubberized palms on some gloves; then you would burn your hand, then you might let go, then you would fall and be seriously injured or more likely killed—then the lawsuits would begin! One company actually put a disclaimer on their gloves warning you to always rappel slowly for the reasons stated above. I had bought a pair of these gloves for $100, and even during a very slow rappel, the palm material was ruined; I was peeved. The mountains and various circumstances dictate how most things are going to go. The equipment needs to be as bombproof as it can be; this is mountaineering, sometimes you have to move out of the way right now or something bad will happen to you. We have technology and brains on our side; there are very few issues that we cannot address and find a reasonable solution for. In contrast, my Kevlar-palmed gloves have thousands of feet of rappels (fast and slow) on them, and the material looks brand-new! I gave the "hot violet" pair to a climbing maniac—my buddy Keith, or El Destructo! After two seasons (roughly), he had destroyed the shells through hard use, but the palms live on. Yes, I know it's a "sue your butt off" happy world, and because one person dies using this type of gear, this can shut a company down for good—whether the gear was at fault or more likely the "bomber perfectly functioning" gear was misused in some way, it happens. We need to realize that even the best gear has its limitations, just like each individual does any one day of the week. Common sense, sound judgment, and appropriate skills have to come into play somewhere along the way. A solo climber free of even his or her chalk bag and sticky rubber can blame no one when they fail, which potentially means they are very injured or dead. Their family in this day and age could probably win a case against one government agency or another for not intervening in some bureaucratic way by warning the ascender of the danger of their pursuit. This idea also applies to the basic free climber layered up with protective gear and ropes. This gear and the lessons to learn to use it properly cost a lot of money; the end user must be solely responsible for the pathway they choose to follow, not somebody else. We live in an information-rich world—

if you don't take the time to find out about the particulars of your pursuit, then just keep pointing your fingers in the mirror, get it! No park service can accurately predict that you will or will not get hit with a lightning bolt or stung by a bee or twist an ankle or some naughty squirrel will rip off your vittles. We take huge responsibilities just by obtaining a driver's license and driving cars. There are so many freakin' variables on the road (i.e., the weather, the road itself, our mood and attitude, plus the opposing drivers). Then, we drink and drive, go too fast and use telephones—all while we are supposedly looking out for the other guy, now who do we blame? Managing people has never been more complex; Moses by comparison had it easy! We all know that the probabilities of danger exist, so we should learn to make prudent decisions, and if per say Mother Nature gets you anyway, don't blame somebody else. Most companies (no matter which subject) with their R&D folks and an army of testers—in-house and in the field—do their very best to ensure that equipment provided will serve us all well. But again, all things have limitations, and we should have the brains and a responsibility to cover our own bases.

By now, you may have figured out that I am an electrician—and just because I like harping, I'm gonna say this: nobody has to tell me more than once that electricity is dangerous, I accept it and became qualified to deal with it in a safe manner, and when forced to do live work, I understand the consequences of a mistake or some other failure that would compromise me or somebody else. Here's a disclaimer for you: Don't try writing a book after you've had one million volts run through your brain, 'cause it will fry your brain, so who can I sue?

Okay, back on track!

The next disclaimer I came across was with crampons, which are the spikes that climbers attach to their boots to ascend snow and ice! The warning stated that if this crampon was used in very cold weather, the toe bail (the part that grips the front edge of a boot) might break! Of course the connotation was, that you would probably fall then be seriously injured or die, and the offending company would get sued, so word to the wise! It's shocking that a top-class company would need to issue a warning like that especially on this type of equipment! When you are using crampons, you are not on some beach in Florida impressing the natives, but are most likely freezing your nunnies off on some freakin' freezing-cold mountain or ice flow. This kind of juju just wraps me around the axle (another way of saying it makes me mad). To the companies, I say: Make sure your equipment is well made and well tested, and if you can't design it so it can perform its function in its intended environment, then leave it to someone else who can; you owe your customers this much. To the user, it's the money-hungry unskilled crybaby weenies that scared a great company enough to have to write a disclaimer like this in the first place. That isn't to say that the manufacturers are perfect, they are not, but most of them are trying to produce good products and would like to stay in business too. It is unusual for top-end equipment to fail unless abused in a very unreasonable manner, so as I said before, be freakin' responsible. Otherwise, I guess the best thing for us all to do is to just stay home and die on the couch.

While I was involved with the glove market, we worked hard to perfect them, and it may well happen in good time; however, by that time happens, ice climbing and mountaineering may be against the law because of a bunch of pantywaists who should have never ever touched an ice tool.

> There are some rules in life that never change,
> Rain falls like the rock and the ice,
> Equipment fails just like skill and muscle and memory.
> Logic and reason disappear with the lightning that just touched down.
> It's a harsh world that cannot be completely tamed,
> Predictability is itself to a wise soul, unpredictable…

Not too bad for a guy with a loose seam in his head, eh!

Another area of the market I worked pretty hard in was the flashlight and headlamp industry! I was quite interested in these light sources because as a military guy and as a climber and an electrician and a home owner! I used flashlights and headlamps literally just about every day of the week all year long. Interestingly, as technology has been racing along, creative minds have been digesting and working on the pluses and failings of the incandescent type of light. The typical flashlight/headlamp used the most basic of all incandescent bulbs backed up with a general purpose reflector and ran on one version or another of alkaline battery, and they did pretty good, and with them, you could at least see what you were doing.

Then some smart people created krypton-star and halogen-type bulbs and produced better focused reflectors, and the lights did quite a bit better! However, this era of lights had two major malfunctions; they used up their batteries pretty rapidly, and the bulbs burned up after a short amount of run time. Turning the light on and off decreases the bulb life too and is known as strike time. I had noticed that the shorter run time and bulb life was mostly associated with the higher-intensity bulbs like halogen and krypton.

All other things being equal, the bulb in a mini Mag flashlight, which was a krypton star, lasted only twenty hours, but, man, was it bright, and the beam was pretty clean, and it was focusable. This little flashlight revolutionized the lighting world and for a very long time. It's safe to say that just about everybody has owned or been issued one of these little gems! The higher-intensity flashlights really helped everyone function better in their job—be you an electrician, airplane mechanic, cop, plumber, firefighter, etc. But the problems remained—despite the brightness and great beam length and color rendition, the user better have some spare batteries around and at least one spare bulb, or it could be lights out at a most inconvenient time!

One day almost two decades ago, Wayne Gregory, my friend and my mentor, showed me a little keychain light; it was quite small, but the little LED (light-emitting diode) that was powered by a small watch battery was effective at helping you see a lock for your house or car in the dark.

Wayne told me that this was the future in lighting and that he wanted me to explore the subject and maybe produce something useful for the outdoor industry. Wayne also mentioned that I should have no problem doing this because I was an electrician. *Jeez*, I thought, *I wire and troubleshoot buildings!* I did have some electronic background even though I was a commercial electrician by trade. I also unfortunately owned a creative section in the sponge between my ears! My initial focus was to integrate LEDs into headlamps for the purpose of climbing and backpacking. So I contacted Hewlett Packard's R&D department to see if they could help me obtain the LED component and its associated data sheet. The engineer for that department asked me what the potential market might be for an LED light source, so I took a wild guess and told them that if my idea panned out, it could mean to the LED supplier literally hundreds of thousands of required product within the current technology. The most powerful LED at the time that HP (Hewlett-Packard) produced was a "sun power" diode and was being used in crosswalks and stoplights; if you think about it, they are and were easy to see even blocks away! The engineer told me that these little LEDs were capable of doing eye damage if you looked directly into the beam, or I guess I could say bulb! HP, free of charge, sent me about a hundred of these LEDs in a variety of colors, which was limited to red, green, amber, and blue at that time and several view angles. Incidentally, the view angle is the angle the light leaving the diode reaches at one foot distance. Realize that these LEDs I was experimenting with were meant to be visible not necessarily to illuminate—their design controlled this aspect. I realized this as

I learned about these components in my lab, which means on my kitchen table when the wife wasn't around! I discussed circuitry and power sources with several engineers and created a model that would give up some useful data for Wayne to channel over to a company that he mentored on this project known as LED Lights (said as lead lights); this company was run by a guy named Gary Burningham. During the next Outdoor Retailers Show, Wayne handed me a four-LED headlamp prototype to look at, which I promptly broke, ticking off Wayne and Gary. Feeling sort of justified, I complained that the material was obviously too weak, so they lectured me about prototype material not being production strength. *Oh*, I mused. LED Lights sold quite a few of these lights while on the market but then faded into obscurity within this very aggressive adolescent field. Meanwhile, I had been on an ice climb during the night when my headlamp, for no apparent reason, shut off, which put me in a nasty predicament as there was little to no ambient light! That one experience taught me to have an alternative light source ultrahandy. As a result and despite the extra light source I carried, I installed a LED into the body of the offending headlamp—my Petzl Zoom! The Zoom was one of the great headlamps of its time and was known for superior durability, decent run time, and reliability—just not this night hundreds of feet up a mountain.

I ran the LED off the same battery source but gave it its own control switch; this idea caught on in the market, and many headlamps these days offer incandescent and LED combinations as well as color options. Another headlamp I use is made by Rayovac; it has a white and red LED that run independent of each other and is quite useful for astronomy. I use the red LED for protecting my night vision and the white for cleaning up the scope sight when the viewing session is over.

The infancy of the LED technology literally exploded just a few short years after Gary introduced his light to the marketplace;

if he wasn't the first, he was one of the first! We realized that this type of light was good; in fact, it performed as good as most standard incandescent bulbs, and the LED would never burn out as they were purported to have fifty-thousand-hour life spans. Hmmm…maybe? But to really capture the lightning market, the LED was going to have to surpass the illumination performance of the high-intensity bulbs that we talked about before, plus maintain and/or improve battery run times. The lighting color (rated in Kelvin) was also a problem as it was limited to colors that were limited in their usefulness (i.e., blue, green, red, and amber).

The blue LED became the standard everyday color and wound up becoming a little less blue as a necessity and still exists in a tone that is a low Kelvin rating in the blue spectrum. Realistically this color wound up being accepted by the consumer as status quo when in fact it needed to be closer to the 4,100 Kelvins or what you may perceive as a whiter light generally. The reason is that the color blue tends to wash out trail detail making it look flat, thus easier to step off rocks, miss tree roots, potholes, etc.; it kills detail in these environments! It can also be bad for cyclists and motorists, causing you to not see potholes etc. until it is too late. Also, this color affects the colors of the wire I see in an electrical panel, interacting in a way that makes them appear to be a different color than what they actually are, which could create a dangerous situation for the electrician. As a result, many companies—either because of my data on the subject or on their own—have upgraded to a true white LED (about five thousand Kelvins or a little lower) helping the end user be more effective and safer. Of course you must match the color to the application. Think about driving and shooting glasses, which are generally amber to help bring out detail. Just a note: if you have fluorescent light fixtures in your home, you can get pretty creative these days by simply raising or lowering the Kelvin rating of the lamp you use (you'll find its Kelvin rating on the bulb). I have corrected many lighting issues using this tactic. One example was the ID card camera and computer program at the Utah Army National Guard. It seems that the system couldn't adjust enough to get good ID pictures of darker-complexioned people, so I increased the Kelvins from 3,400 to 4,100 (warmer to colder, or from browner to whiter)! Afterward, the system worked great for all complexions—this beat adding light fixtures in a place that didn't have the space. Unfortunately, the LED technology has always had several negative by-products working against it, and one is in the form of heat! Just like with anything else electronic, the technology is great, but sometimes we march ahead too fast and don't even notice the fallout of the next great thing we have created. I had a LED headlamp fail due to its internal high temperature, leaving me stranded in an attic space that was otherwise pitch-black; fortunately, it cooled off and resumed its normal function, and life was good again.

Another great thing I got to work on relative to LEDs was creating illuminated wand markers for mountaineers! You may recall the disaster of 1996 on Mount Everest where eight people died; some simply could not find their way back to camp hunkering down in a fierce storm! As a result, an electrical engineer (Robert) who owned a small outdoor accessories company had created a LED light that could be firmly installed into the end of a route-marking wand, which is used to identify dangerous areas, caches, and ultimately your way! The idea was that in the dark or during a storm or whiteout, the light might be the thing that brings you home alive! My job was to see how they worked in a total whiteout! Well, you can't just order up a whiteout in Utah! However, I got real lucky, and we had weather in the local mountains that produced my needed whiteout, which, by the way, is a situation where you cannot tell the ground from the sky; it all looks the same, which historically has been responsible for climbers walking right over an edge or not recognizing other similar dangers soon enough etc.

What I discovered was with the LED installed in the wand pointing straight up, I lost sight of it after I was maybe about forty to fifty feet away, the whiteout notwithstanding. So I turned the LED facing me, and just like a crosswalk light etc., I could see this blinking red light hundreds of yards away—maybe we could do some good for climbers and other adventurers after all! Along these same lines, I suggested that the makers of tents and shell gear add reflective material to conspicuous locations of the said gear—it just may save a life too. Finally, LEDs have been engineered to become part of all of the lighting systems for homes, offices, perimeter and street lightning and are getting brighter and brighter and more efficient all the time. I stand amazed at the great minds that have nurtured the humble LED into its current capabilities and am very glad to be alive to have seen it happen. I also have had the opportunity to get involved with water purification and the portable methods utilized by adventurers all over the globe!

At one time, backpackers etc. either drank straight from the stream, lake, or spring, including eating snow, which is not too efficient; we hoped that the water looking clear as a bell was not hiding some issue. Three decades ago, we didn't think much about it; if we did, we might treat the water by boiling it or adding iodine or Clorox, the latter two only helping with some of the issues one may encounter. As the outdoors life became more and more the thing to do, people began contracting and getting sick from things like giardia and dysentery plus a host of other water-born issues, including things caused by general camp uncleanliness! I heard one time that giardia makes you feel like a truckload of dead dogs, so I didn't ever want to contract it if I could help it, but I did contract amoebic dysentery while in South America, and I can say I didn't feel much better!

These days, outdoors people or anybody needing to make their water safe have several options besides boiling:

a. Micro filters that will block just about everything the world has to offer coming from your water source and will even use some additional filtering that employs charcoal and iodine to address the rest of the things you may regret ingesting! These types of systems are typically small and lightweight and are not intended unto themselves as a long-term survival means for creating safe water; they utilize a pump that pushes water through a filter or series of filters into youur storage container and happily—if they don't become plugged—can purify hundreds of gallons of water.
b. There are gravity systems that simply use gravity to push water through a filter into a storage container. This is a slower method than the pump system, which gives you a quart of H2O every one and a half minutes thereabouts.
c. Advanced systems are using ultraviolet light and claim that this method nukes or kills everything that is present and are very simple to use; all you do is scoop up a liter of water, put the ultraviolet light tube into the water so its safety sensors touch the liquid, turn on the unit, and in about a minute, you have safe water to drink! But it requires batteries.

    One of the first companies to use the ultraviolet light technology was Aqua Star! They built a system into the cap of a large-mouth one-quart Nalgene-type bottle that contained the electronic circuitry, the batteries, and the light tube itself; it was pretty compact and ingenious. This method worked pretty good but still presented a few exposure issues that needed addressing!

    For example: Because you were obligated to scoop water into the bottle from the source then purify it, then drink from the bottle, placing your mouth on the rim that had been in the water—the area inside the bottle was safe, but the outside of the container was still a risk! As this technology evolved, systems were created that contained a bottle with a cap with an opening in the center that allowed you to scoop water into the vessel as usual. After the container was filled, you would then place a pen-style ultraviolet purifier through the opening of the bottle cap to purify the water. Finally you would remove the cap and drink safely from the container, supposedly! This process is still evolving! The UV (ultraviolet) purifiers do work and are great alternatives to the pump-and-filter styles available that is as long as you handle them properly and your batteries ain't dead!

d. The military and other users utilize a reverse osmosis method to clean the water and are very effective pretty much no matter where you are getting your water.

All these systems work very well as I have used water from all of them from questionable water sources and have never gotten sick as a result.

Lastly, it is simply unwise to drink unpurified water and see how it goes.

Speaking of evolution, batteries are another technology that is part of our ever-changing world, which is becoming more and more reliant on technology and electronics to survive. Surprisingly, companies have moved forward very quickly with the idea of powering headlamps and purifiers and a dozen other things with fixed in-place L-ion batteries that need to be recharged often and did not offer an alternative "just in case" means for power. Granted the run time and performance of these batteries relative to the device, the power can be quite long indeed, but there is always going to be a day that things go different than what some engineer planned on from the comfort of his lab! I'm a field repairable type of guy, and in this day and age, many things are of the nature of—when they are dead, they are dead—even for McGyver!

Recently, I tested a LED headlamp that got very hot (126°F) on the housing that held the LED and happened to be the part that you would touch with bare fingers to adjust the light angle; its battery was fixed and, on the high mode, only lasted several hours! So if you didn't have a charging method handy halfway up a mountain, life could become bleak—happily the light did have lower light level modes, which you could use to sustain battery life up to about six hours! But in the end, a limitation had been introduced, and what if you had forgot to check the batteries' state of charge before you left home, now what? I also tested an ultraviolet light purifier using the same fixed L-ion battery idea and found that after about thirty discharges, the unit shut down and required a reboot. So if you didn't have a means—meaning a USB cable and a charger to reboot its computer, you were finished, show's over, and kaput!

I will say this though: the companies I have tested for have worked very hard to make things simpler and more reliable than ever, and these two types of products have developed into some of the most reliable, cutting-edge gear money can buy! I am a real fanatic about the reliable and durable nature of my equipment, and I use these two products to this day, and they have proven themselves time and again very worthy. From somewhere around the midnineties to present day, I have worked hard in the outdoor industry as a field tester and have held my own against in-house testing staff and organized definite purpose-testing companies that wound up adding me to their efforts off and on—it's felt great knowing somebody valued my opinion.

Often I would bring my family and a few friends to the trade shows to see what gives. Most of the time, just my son and wife came since my three daughters had better things to do; Wayne usually helped us get in. My wife enjoyed wandering around the show and seeing all the cool apparel and so on that would be available for ladies and got to meet some of the folks I worked for—everybody liked her a lot and told me how lucky I was to be accompanied in life by this woman; I couldn't have agreed more! My son Eric would come and help me and the rest of the Gregory gang set up the booth, then he would wander the show scamming free booty along the way; he had also become a microbiologist and helped run a local laboratory. During one show, a company who knew I did research for water purification was showing me their prototype; I realized that their needs were right down Eric's alley so to speak, so I told them about Eric and the lab he works for and got them together for the needed detailed testing!

Another time, during the setup of the summer OR Show of 1999, a category 1 tornado hit Salt Lake City! It was around 11:00 a.m. on August 11, and there were roughly 4,500 booths being set up in the Salt Palace, with a few overflow tents erected in the parking north of the main building. The whole setup operation involves hundreds of people and dozens of forklifts and aerial platform lifts all in motion simultaneously. It's a busy place; the show starts in just a few days, and you gotta be ready so the retailers can come to your booth to see your wares. Eric and I were in Gregory's booth working on a lighting system when suddenly, we heard lots of people cry out in a crescendo, indicating something or someone was falling; then we heard a crash. Seconds later, all the lights went out; then the emergency lights came on. So including the skylight panes, we could at least see our surroundings! In the next moment, people were in a total panic running toward us from the north end of the big room we were in, screaming, "Run, run!"

Eric, in a concerned tone, yelled at me, "What do we do?"

"Nothing," I commanded. I didn't want to get into the stampede. We didn't know yet what they were running from or if it was safe to follow them! In the next booth was a pregnant lady who was very alarmed; we kept her with us to protect her from the charging crowd; she would have been trampled. A minute or two later, a huge roll-up door to our west, the kind semitrucks drive through, tore open; at that moment, most of our questions were answered! When Eric and I had left home around seven thirty that morning, the sky was clear and the day promised to be very nice and probably hot. Now, through that door, we could see that the wind was horrific—and I mean horrific. Visibility was very low due to dust and all other kinds of debris flying through the air; we couldn't believe it! Then somebody over by the east main entrance yelled, "Evacuate the building, there is a gas leak!" We were ushered out the main entrance and told to take refuge in a hotel across the street! When we made it outside, the scene was unbelievable: trees all along the street known as West Temple were destroyed; they had been uprooted, snapped in two, or were gone to who knows where. Garbage was all over the place, and the feeling of panic was ever present with police officers guiding people to safety. We were shuffled into the hotel lobby and was advised that we had been hit with a tornado…in Salt Lake freakin' City—nothing like that happens here; after all, the Temple is just a block away with Moroni and his horn standing watch! Unfortunately, the tents used for the overflow that was on the parking lot north of the Salt Palace were in the direct pathway of the tornado and suffered significant damage. Many people who were inside these tents were injured, and sadly, one man was killed!

The WyndHam Hotel, which sits between these tents and the Salt Palace, was where Wayne Gregory and George Marks (a famous tent maker) was talking business; their room window faced south. Suddenly, Wayne, to his horror, sees the tornado about to hit them, grabs George, throws their door open, and they dive for the hallway! In the next instance, the room explodes as the tornado crashes head-on into the structure; moments later, the tents are destroyed. People around the valley saw several of what turned out to be electrical explosions, as the tornado waded through that end of Salt Lake Valley, one of them causing the Salt Palace to lose power. The city received quite a bit of damage, and us Utahans realized that not just the Midwest can have a serious event where wind is concerned. What amazed me with the Show folks was how everyone pulled together and helped one another out by loaning booth space to those whose booths had been destroyed. Wayne invited a handful of companies to set up skeleton displays and conduct their business right in his booth. The sad thing was that someone had died during the storm; however, the feeling of genuine caring and concern for others' welfare was so ever present during the rest of this otherwise successful Outdoor Retailer Show!

During the 1990s, it seemed as if I kinda got reborn and received some payback for the lightning strike! Much of this all seemed to heavily revolve around the trade show. Becoming affiliated with these folks seemed to help fuel my need to live; they helped to serve my need by letting me serve theirs! I also made quite a few friends during this decade and got to learn how to Telemark ski, snowboard, target shoot, and even become a better climber; I really got to taste life again. All of this was despite the fact that I was still being repaired for one thing or another; little did I know that I would eventually have to give up many of these fun things just for the sake of survival and quality of life.

I had been testing and tweaking packs for Wayne for a while now and was beginning to see some of our collective ideas become products for the backpacker! One summer show, I noticed a pack that was sitting against a wall in the Gregory Booth conference room. It blew me away! This pack, the Acadia, was the physical in-the-flesh culmination (inside and out) of what we had been working toward for several years, and I had to have it or one like it anyway! Dion, one of Wayne's managers, seeing my excitement, had to burst my bubble by telling me that this pack was headed to a show in Europe and was intended for that market! I flipped out and started ballin'; I even spun around on the ground a bit and just about went into a Blue Funk! Finally, Dion and Wayne said I could have that pack, but I had to produce a very involved test report that they could use for their customers in Europe—and that I only had a week to get it done; happy as hell, I said, "No problem!" So my son Eric and I banded together and headed for the Wind Rivers, a beautiful rugged mountain range in central Wyoming! As it so happens, everything we had with us on that trip was being tested except our undershorts. On this trip, I was testing a new line of boots for Asolo, which was known as the FS line. We definitely did some good for the company identifying a problem this footwear had; for example, up in the toe area, the boot would create blisters on top of your toes, which really became a problem during the forty-mile march through this supremely beautiful landscape. I tried several different sock combinations, but the damage was done, and moleskin became my best friend! I was also testing a headlamp that broke as a result of the wrong material being used. I also tested a lightning detector for a company called Strike Alert, which would detect and track lightning as far away as forty miles. Before we left for Wyoming, I asked the company just how far I needed to go with this test because I had already gone all the way with lightning once. They told me that I didn't need to go that far again. Nodding, I thought, sure, okay! A few days into the trip, while up in the tundra around Reid Lake (around 10,000 elevation), we spied a storm approaching from the northwest; the detector was very accurately detailing that aspect and became very active when the storm was twelve miles out, which the company and the detector says is the death zone where lightning is concerned! Pretty soon the storm was right on top of us with the device pointing its electronic fingers (blinking red LEDs) at me! The unit had given us plenty of advanced warning of what was coming, and I can tell you being right under that storm now was pretty dang uncomfortable.

While up in this region, Eric and I caught a bunch of fish, relaxed, and had a pretty good time. But I gotta tell ya, getting to this basin we were in was hard work; it's about a 3,600-foot elevation gain in the first six miles. Fortunately, our loads were light at about thirty-two pounds each, plus we were in pretty good shape, allowing us to make it to our base camp eleven miles from the trailhead in about four and a half hours. It's worth noting that for the first time in all our trips together, Eric led us up the trail the whole way, often peeling my seemingly lifeless body from logs that I had rested on and became part of—man, he could hike! Once, while passing through a meadow near the upper plateau we were headed for, we saw wolf tracks; this can really put you in your place as the human is not always at the top of the food chain, it was pretty cool!

Another example of the food chain occurred one winter in our own Uinta Mountains and just a few miles from the town of Kamas, Utah! Eric and I had snowshoed into a place called Pine Valley; it's where the upper and north fork of the Provo River converge to flow together westward. That night while in our tent, Eric pointed out some small spiders that he claims to "wait for you to go to sleep then crawl up your face and eat your eyes out of their sockets!" Great, I thought, I hope he is wrong—but who am I to argue? He is the resident biologist. After we dealt with the spiders, we blew out the candle lantern, and it was then that we heard the wolves' howl. This was a first for us; wolves weren't supposed to be here, but they were, and a few of them were in the same river bottom a hundred yards away! Eric asked me if I had brought the SIG .45 pistol, I said no.

With eyes growing rounder and rounder, he said, "What's the matter with you?"

(So dramatic is he!) I replied, "You are mad 'cause you ain't the top of the food chain anymore, and you caught it in the butt!"

As we sat there deliberating wolves and eyeball-eating spiders, a big patch of snow fell from the tree we were camped under, hitting our tent and scaring the hell out of both of us! After that, we slept occasionally, with the wolves singing several times more, and the coyotes yapping in chorus. It was actually pretty cool hearing that voice that has been lost in most places to change! As for the spiders, I hated the idea of running around with empty eye sockets! The next morning, we snowshoed around on the frozen river itself and found the tracks of the wolves; they were amazing looking and were as big as our hands! We took a few photos for the Fish and Game and enjoyed the rest of our time in the beautiful landscape of Pine Valley.

Anyway, the Acadia pack worked just great, plus we got lots of input for the other companies and enjoyed a trip into one of our country's true wild places, complete with a really intact ecosystem, bonding deeply as friends and as a father and son.

More recently, Wayne and his design team needed some help with a specialized mountaineering assault pack, so Eric and I hiked every trail in the western Uinta Mountains to get the info the Gregory folks needed! I carried forty-five pounds on each of these trips, and we really got into great shape, even met a few mountain goats along the way too! One time, we went down what's called the Christmas Meadows trail. It led us to a place where we could see the cliff I had fallen down after the lightning strike got me; it's a beautiful and sacred place.

I went back to this same area a week later alone, lugging the test pack clear up to a high basin that pretty Kermsuh Lake sits in; it's eight miles one way from the trailhead. This trail is quite rugged and steep and full of bogs, but it's well worth every step; the scenery is outstanding. I remember getting to the lake feeling pretty beaten up, especially my feet, which were hurting pretty bad and very swollen! I didn't have a lot of time to enjoy the area and was going back down straightaway, pushing my limits as usual, traveling fast but not exactly light! I pulled off my boots and soaked my hooves in the cool lake water wishing I was spending the night. I wondered how many days like this were left in me (in my mind, as many as I want), but then there is the reality of my problems. I'll go till I don't want to anymore, I reasoned. I also remember thinking that I better get in my trail time now 'cause I don't remember anybody promising me a certain amount of time while in life. Many people only get to experience a brief life; they die at birth or soon after while others go on forever or cash in midway. There is no set bank account of time for the duration of a specific life, only a hope. Remember this: life begins at point A and ends at point B—whenever that is, so you better get the right things done in between 'cause your time might already be up, friend!

I thought these thoughts while sitting in this very beautiful glacial cirque, resting my feet and my soul; consumed by history and endless life, wild flowers bloomed, the water was soothing, the sun warm. The majesty of the huge 1,2000-foot peaks—A-1, Ostler, Spread Eagle, Agassiz, and Hayden—so protective, surrounded me, made me feel grand and humble, and I knew that I was lucky; my life, no matter what, had been good, and again, I was home.

During the 1990s, I did a fair amount of climbing with Keith, Fred, Tony, Bob, and Eric! But it was Keith and Fred, along with some of their friends, that I spent my most technical adventures with, which also allowed me to help climbing companies create better and safer equipment. One time I was on an ice-climbing road trip with Keith; his wife, Joyce; and a friend of theirs,

Thad, who was a very decent climber! Because of jobs and other constraints like money, Keith and I left late in the afternoon from Salt Lake City, after I had sold a sleeping bag to provide money for my part of the needed funds, with the agreement that he do all the driving if I do all the talking to keep him awake. No problem! We arrived in Ouray, Colorado, around 4:00 a.m., meeting everyone else at a condo that Johnny, one of Keith's friends (an ace rock and ice climber), and his wife had arranged for. With so many people under one roof, I wound up sleeping under a set of stairs in my sleeping bag with Keith's dogs; I liked them a lot, and they kept me really warm.

One day while we were climbing in the Uncompahgre Gorge, which is now an ice-climbing park where water is dripped from an aqueduct forming immense and beautiful ice flows that us ice climbers savor, Keith had to teach me a lesson because I was being a little too cocky! Shortly after Keith and I set off up a flow of ice that contained little stretches of what we call verglass (which can amount to nothing but a smear of ice)—on belay, I started my way across this dicey section; suddenly I exploded from the windowpane thin ice, coming to rest upside down against a big rock. This short little spill really shook me; I had believed I would never fall from the ice—never—but I just did, and what if I had been ropeless…hmmm.

The old mountain guide Keith never said a word; he didn't have to. I collected myself and carefully progressed through the verglass and to the belay. Now I understood what someone meant when they said, "You don't know what you don't know." Just because you think you are good does not mean that that is the reality—it may be very different, and you may learn a harsh lesson in finding out the truth! This lesson made me think about how sure and safe Fred was about everything he did!

One time, while out climbing in Big Cottonwood Canyon, Fred wanted to see what the view was like from the top of a small cliff, so the idea was to skirt around and find a walk up route to the top. As Fred made his way around the cliffs, I just simply soloed the face and met Fred at the top. The look on Fred's face was not what I expected; he was shocked and mentioned the risk I had just taken! I had just made Fred very uncomfortable, even though during normal climbing routines, he was very comfortable with my climbing and rope-management skills; we had formed a tight bond with each other and the rope, and we protected each other with it. I had lots to think about and consider especially with how my actions affected others, especially since I had been given my life back and so many good things were being sent my way.

I remember a close call one spring on a solo snow climb when things were getting ugly. I did manage to make a good decision, which was to go home as opposed to risking what would have been a death fall into a huge, deep snow hole! I had been mountaineering up a deep canyon on the very thin snow and ice of a dying glacier that covered a rushing freezing water flow sixty or more feet below and completely out of sight. Presently I came to an opening that exposed the dangerous truth. I anchored my rope to a nearby rock outcrop, tied in, then carefully crawled out to the opening, turned on my headlamp, and peered in; my heart raced! Where I was lying, the ice/snow was dangerously thin and was hiding a house-sized cavern with a rushing torrent in the bottom; you go in that black frozen hole, you are gone. It simply was not prudent to continue! I went to Keith's house to tell him about the adventure and close call; Joyce (Keith's wife) was listening. After I finished with the details, she looked into my eyes; then in a pleading tone, she said, "Please don't break my heart," then left the room.

I utilized Keith, Tony, and Fred often while testing gear because of their very intelligent perspectives. Fred was amazing in terms of craftsmanship, so was Keith and Tony. Tony at one time was the youngest person to climb Denali and has established routes all over the country and is an author on the subject. Tony has also designed and built safety equipment for climbing, which really impressed Wayne Gregory, who just loves creativity, especially when it's mixed with action that produces something tangible. As I have been reminded over and over, the difference between what is seen as success and failure is often just the taking of the next step—from dream to vision. Wayne wanted a better pack than he could buy. So he bought a sewing machine and, right out of his garage, started turning his desires and dreams into something usable and saleable. Actually for him, the creative roots go back to a merit badge in scouting generated by a mentor—sewing! You can go into any nook or cranny in the outdoor industry and find that most of the companies had similar beginnings to Gregory Mountain Products, and they all were helping me find my new place—amazing.

Many people came into my life that in their own way helped me overcome many of the challenges the accident imposed on me, and each one of them taught me something important. For example: patience and safety as applied to everything you experience. I also relearned attention to detail and how to have fun too! Fred helped me learn how to snowboard; we enjoyed hundreds of nights up at Brighton in Big Cottonwood Canyon, skiing and snowboarding for cheap, using "two for one" passes. Boarding for me was about the most fun thing I ever did, especially on Utah's high and dry powder days! One time, Fred and I did a ski mountaineering tour in the Brighton backcountry. We were trying out an awesome Eureka tent that made short work of the high winds we were exposed to one night. We also learned the value of hanging stoves that we would later help a few manufacturers refine. We skinned up hills with heavy packs; then with our loads dumped, we would ski virgin snow. It is an understatement saying we had a great time! On the flip side, Fred and I had gone climbing at the City of Rocks in Southwestern

Idaho and were having a fun time getting lots of miles of vertical in, comparing gear and eating good too. One morning after breakfast, we went to climb at a place called Parking Lot Rock! Fred asked me to grab the gear from the tent; I replied, "It's in the truck." He corrected me saying, "I was just in the truck, and it's not there, so it must be in the tent." We realized that our truck had been broken into and all of our climbing gear had been stolen, collectively about $2,500 worth, I flipped out! It takes years to acquire what we had plus lots of money!

We went around to other climbing camps telling about our loss, only to find out several others had suffered the same low blow! So Fred and I drove to the ranger station a few miles away to report the incident; then were told to go back to camp where a sheriff and a ranger would come and take our statement. On the way back, we pulled into a camp with several vehicles and four teenage boys hanging out. When we told them the story, their level of concern seemed to me to be very low! Based on my experience from the past working as a security policeman for the Hilton Hotel chain and working (on occasion) with Salt Lake City Vice Squad, I knew it was them. As we drove away, I heard one of them laugh, and I mentioned it to Fred, saying, "I know I can make them talk." He replied, "We can't prove anything." So we ambled back to camp to sit and wait for the authorities and consider the fact that our climbing days were over for a while! In the meantime, these idiots came driving by (criminals are stupid) and went right down into Parking Lot Rock to do some climbing with, I am sure, all their new (*our*) gear. Fred told me that he had a funny feeling about these guys now and to stay put while he goes and has a look-see. A few minutes later, here comes Fred at a run saying, "Get into the truck!" We both jumped in; then he directed me down into the parking lot where their minitruck was parked and told me to ram it with my 1989 4Runner!

I was like, "I'm not gonna ram that stinkin' truck."

"Well, then pull right up to it so they can't escape when they return," he urged, so I did that much. Then he told me to look in the bed of the truck. Wow, sure enough, there was his camera! I know it was his 'cause it had eleven pictures left on the amount window (it was a paper-film travel camera) and had a dent in one corner that got there because I dropped it off a climb the other day, incurring Fred's wrath. Seemingly, Fred softened up a little; then I got mad, saying I was gonna knife all their tires! Fred waned saying, "We can't deny that they may have simply *found* the camera, they would say that anyway." So it was decided that Fred would wait at the vehicle and I would run to the ranger station for proper help! Along the way, I alerted other climbers who sent a one-hundred-person strong contingency to back up Fred, while some others gave me a lift to the ranger station. The suspect climbers returned to their vehicle with quite a reception waiting, including a sheriff, a climbing ranger named Maura, and a mean-looking crowd of climbers. The ranger was pretty smart; she first asked the thieves to unload their packs. These guys had enough gear to climb the biggest aid routes known to man, or in other words, way too much for typical free-climbing of standard routes, which was decided by way of consensus by the crowd of edgy climbers!

Then she asked each suspect individually where the camera came from and got a different answer each time. She asked them to let her examine their camp and would get a warrant if necessary, and she immediately found several items that Fred and I had clued her in on to look for. She then had the suspects brought to the camp, read them their rights, and handcuffed them, then she brought us down to identify our gear. Fred was supposed to keep a leash on me because I was causing a stir back at the parking lot. Sure enough, before he could contain me, I had spotted one of my climbing boots lying by my Chouinard pack, bolting from the truck, thinking I felt Fred's fingers rake my back, I was gone.

I grabbed one of my boots as I beelined it for the handcuffed creeps, managing to get a great swing with my boot to the side of the head of the kid we called Scruffy Jaws, then following through catching two others with a little boot as well. Next thing I knew, I was wrestled and detained by the law and was threatened that if I didn't behave, I would lose my climbing gear forever. Fred was giving me the evil eye real hard, so I had to settle down. When court time came, Fred said I wasn't going to be allowed to come, so he went and had to stare down those stinkin' animals with the prosecutor! Happily, while out on bail, these idiots went to another major rock-climbing area in Idaho and did the same thing, but were caught by the climbers and were detained climber style before the law arrived! It seems the climbers roughed them up a little then stripped them naked, tied climbing ropes to their ankles, and hoisted them upside down up the rock about thirty feet, and tied them off to remain there till help arrived. Well, the climbers wound up in trouble for doing justice! In our case, these kids got off paying some restitution and doing a little time, and we got our gear back, as did everyone else who had been stolen from! I guess one other plus that happened was that a fairly large fencing operation was shut down as a result. I gotta tell ya something though: I have a problem with people being defended when they were caught red-handed; there's nothing to defend. How is this justified?

Tony and I made it to the City of Rocks several times too. Tony had me come along on several occasions to help put up new routes and assess the difficulty from my skill level perspective, which is a bit lower than his. Also in his own unique way, he worked me through some of my little isms that held me back as a climber, plus he always gave very valuable insight concerning whatever manner of gear I may be testing at the time—be it a harness, a pack, or climbing boot; he was quite the technician. One time while ascending a route that would get us to the route we were going to put up, Tony, his friend James, and I simo-soloed up several long pitches, which means you are tied together with ropes but use no protection; it also means that if one guy slips and falls, you all fall—what a great amount of trust that involves. I had to be rock solid that day with those two guys who were far better climbers than me and far faster; they flew up that stone, and I flew with them—it was such a great time. Another time, Tony and I, with my family, traveled to the City of Rocks for a very long day of hiking and climbing; when you climb with someone with so much skill, even one day can add lots of fortitude to your capabilities. They can see the details of what you need to work on and help you overcome it. Tony told me one time that he climbs hundreds of routes through a summer, which says a lot about how familiar he became with routes all over the country and why he was so confident; time and effort and proper instruction are some big keys to success. I think back to a kid named Zack who I ran into routinely while soloing the Great White Icicle in Little Cottonwood Canyon. We wanted to climb together, we even exchanged phone numbers, helped each other with specific info about other climbing areas; the chemistry was good, but we never hooked up. I think we were both afraid to let our guard down, which is sad, and I blame myself 'cause I know from our encounters up high, that it would have been fun. Maybe we just needed to develop some trust, then all we had to do was tie the rope between us and head up. Sorry, Zack, maybe if we are lucky, we'll meet in the next life and get another chance to realize the brotherhood of the rope.

Something I realized along the way was that I had climbed with some really great climbers, but when it came to certain techniques, each one of these guys did it exactly the same way; and at that point in life, their paths had never crossed. It cemented in me the fact that some things are always wrong or right as the case may be. Amazingly these guys were the ones I turned down all too often when it came to getting out and spreading my wings and becoming better at our craft in the mountains. I realized that I did the same thing in other facets of my life too; I simply wouldn't let people get too close! My closet of skeletons was obviously quite controlling and full of doubt and represented a giant hurdle in my life. I wonder how I got this far, and for that matter—anywhere. The limitations of being alone and enclosed can be so monumental!

But then there are the examples of others enduring life's hardships, alone or imprisoned in some way, that keep you inspired. I didn't know it then 'cause I was too busy having fun, getting repaired, and trying to forget; but the lessons, be it subtle or direct, that I was learning through this time was preparing me for my next phase of life!

I am sure that somebody was watching to see how well I would do with my new and existing circumstances.

If I haven't learned anything else along the path of my life, the all-encompassing Boy Scout motto, Be Prepared, would become one of several lifelines to save me over and again. Change, perception, and opinion, along with paradigms, habits, and the weather of nature and people, including resourcefulness—I've learned that these are great things to understand! Because fires and wars and rumors of many sorts were coming!

14

The Final Battle, Trial by Fire, Another Warning, and the Rat

I had said earlier that I was a very busy field tester, especially since it was directly tied to all my outdoor pursuits (i.e., climbing, skiing, snowboarding, snowshoeing, hiking, and backpacking, cycling and astronomy), I was busy any way! These activities provided the fuel for testing, and they were also part of the fuel to my being alive, and it all required precious time and money. I was also giving regular lightning awareness classes to the public. Interestingly I didn't advertise; people just seemed to find me, and in a roundabout way, and we would wind up on the subject of lightning. Most often there would be a need, so I filled it at no cost to the people desiring to hear my story and advice; it was my duty to help those people! Also, I would forever have to manage my injuries from the lightning bolt, and was ever vigilant in regard to my health and physical condition. I needed to remain in the mainstream of life, including serving my country as a member of the Air National Guard, with whom I served from 1975 to 2008.

As you may recall, the unit I was in (151st Civil Engineers) was a construction unit and never turned down an opportunity to serve. I deployed fifty-two times in my career; I saw it as an honor and a privilege—always. The deployments were usually relatively short (i.e., two to three weeks in duration) but very high OP (operations) tempo affairs! Sometimes, however, we would get assigned to go out for six months or longer to places like Iraq or Afghanistan when those conflicts reared their ugly heads. In the military, each career field has its own set of requirements relative to training, which requires you to attend formal schooling and subsequent correspondence courses including on-the-job (OJT) training done within the unit you are assigned to. Some CDCs require a month of after-duty hours study to complete while others require up to a year of constant effort. Some of this training depends on your unit's mission and is usually specific to a career field! The wars or campaigns took many of us away for months or years at a time, and some of us over and over. I finished my civilian service to the National Guard as their lead electrician, having been one of their full-time sparkies for twenty-four years. I had begun this part of my civilian Guard career in 1989 and retired on January 15, 2013, serving the military effort for thirty-eight years overall! I often joked that if the enemy took me prisoner, they would play hell trying to figure out who I really belonged to. Was I a civilian, or am I military, and was I army, or am I Air Force? Who really knows? I sure as heck don't. They would probably send me back crying foul.

Anytime you are involved with the military, the top concerns are the mission, and are your people ready to respond, a.k.a. the readiness level. During my experience with the National Guard, I was always very busy, whether I was in the uniform or working as a civilian; we were minutemen! It was always a very time-consuming thing to be at a high state of readiness because I had my own military shop of airmen to train, who were a mix of various backgrounds within their civilian occupations (i.e. schoolteachers, truck drivers, janitors, even bankers). They needed to be capable of maintaining and constructing electrical systems on our base and abroad. Plus, we had wartime tasks to train for too, which involved being able to do our job in a war arena even while wearing a chemical suit in blistering-hot or -freezing temperatures. Some of our training required us to team up with our active-duty counterparts and learn how to homogenize different branches of the service into an effective team. We had to understand the flow of the mission, and each man or woman had to be ready to insert into any situation as needed; it's asking a lot, especially for the citizen soldier. We also had to be qualified to carry and shoot several different types of weapons because we just might have to defend ourselves while constructing a runway or building a tent city! I was also responsible for constructing electrical systems for other units on base—to include maintaining their existing systems, and I routinely had to keep myself moving forward in my own upgrade training to remain effective. I also performed other duties, including making sure everyone within the shop had a current mobility packet, which contains all sorts of things—from pay status and living will info to dog tags and shot records, and that they were up to date. Fortunately, this one thing was such a big deal that our unit provided routine inspections for these items, including chemical suit inspections and weapons qualification to help keep us on our toes. Training and readiness, training and readiness—we ate it, drank it, and in our sleep, dreamt about it! While at the Air Guard as a military and civilian guy, I was exposed to a very highly organized process; our sense of order was anal and would drive many people to drink. The Air Force has a form and exact process for everything, even going potty! We may have appeared to be robots in some eyes! However, we were allowed and expected to think for ourselves and were taught teamwork and independence simultaneously also.

The red tape at first seemed redundant and a waste, but it left little to question and kept us very organized, allowing us to move forward intelligently and it never bogged us down! I found it to be a great way to attend to and manage so many different requirements, including people and situations, and was pretty simple to apply. As an example, my 623 training record was about the size of standard paper (i.e., 8.5" x 11"), and at the end of my career, it amassed the thickness of maybe one inch, yet it contained every bit of info relative to who I am as an Air Force–type brat, including all the training I had received over all those thirty-two and a half years in the uniform; it was a log of my military history, including deployments, counseling sessions, injuries, etc. It represented my entire life in the service! Our unit was well-known and respected for our can-do attitude; we hit the ground running and refuse to lose. Overall, this was a great environment to recuperate from the effects of the lightning bolt because I was going to get long-term help that would do me a lot of good.

In the year 2000, I was thinking about moving over to the Army National Guard side of the house, down on the south end of Salt Lake Valley. My friend Tom was the civilian maintenance and construction foreman there and said he really needed my help. Lately I was becoming impatient with my boss. I thought he was too self-serving and was too narrow-minded; also, our top leaders were becoming of a mind to limit or stop upgrade training. The opinion was that especially with us licensed electricians, we had got our training decades ago and didn't need any more; of course, this is outright foolishness. I also remembered being forced to work out in the snow after one of my neuroma operations. The affected area was not supposed to get wet, but it did, causing me a longer recuperation and increased some of the damage; I was supposed to be on light duty, but was told to put a plastic bag over my foot if I was worried about it.

Something else that irritated me was I had been assigned to get control of a problem at an army camp in Utah's west desert, known as Dugway Proving Grounds. Our general had issued an order to my boss to get an electrician out there to make sure the site was complete and ready for a major exercise that would be happening within three days; I had my marching orders. I went, I exercised great amounts of skill and authority that tough and ornery old master sergeants are famous for, completed the mission on time as ordered, and met someone who would become one of my best friends in life, and then I came back home—happy days! Later on, a ceremony was held in honor of the people who had supported this exercise in the desert, which involved thousands of personnel from all over the world; I was never even mentioned—nobody said thanks or kiss our rosy "you know what." Bear in mind, I did not get into the military to receive accolades and can live without them. But I was told that my boss got yelled at by his boss for sending me out; I reminded him that our general was the requestor. No matter, I was told to forget all about it—you are not getting recognized! I was sore at these people because of the principle of it all; their heads were so big these days that even a general's order seemed to matter not! Mark and I had the Air Guard base in spic-and-span shape electrically. In fact, it was running so well that we had not had any after-hour callouts for trouble in more than a decade; this place could just about run itself. Again, my need for challenge was needling me. I had worked very hard as an electrician, a climber, a tester, and a family man throughout the last decade and had moved a great distance away from most of my injuries that were a result of the lightning strike, but I needed to go further! I was burning inside to increase that distance, so hard work with a positive focus, to me, was the key.

I was vacillating back and forth trying to decide what the best thing to do was. I had noticed deep inside me, despite my gripes, an instinct telling me to stay where I was, that if I went over to the army, I was headed for fire and that I may even get killed there; I could sense the danger! Before I decided to decide, I called my friend and now longtime mentor Wayne Gregory who lives out in California! I explained all the pros and cons about leaving the Air Guard and heading down to this place that neither one of us knew anything about! I asked Wayne if he would pray for me because I felt that he was very close to the Creator; he said he would call me back soon with his thoughts on the matter. A few days went by; then I got a call from Wayne. He told me that I should not take the job, that I would be exposed to more negative situations than I may be able to survive. Wayne knew I was strong, but this was different, he warned. Despite this warning and my own instincts, including my commander and chief master sergeant telling me how disorganized and poorly funded the army was and how old the facilities themselves were, that I would be back on my knees begging for mercy, I went anyway.

Now, you may have also noticed that I am still at this time undergoing semi-frequent operations due to the fallout from the lightning strike and was well beyond the state of being too busy, so much so that there was little time for sleep. I was holding down full-time and part-time positions with the military, deploying one to two times a year, going to school off and on, and wading through career development courses relative to my military and civilian career field, plus I was climbing and field testing too. Wow, I need some air! There it is, a great recipe for stress, and a lot of people do this very thing as a day-to-day routine until it wrecks things beyond repair. Somewhere in that last paragraph, my family still existed too. Believe it or not, I managed to spend quite a bit of semi-quality time with them, making it to ball games, dance and violin recitals, even going camping or on family vacations to our local mountains—the Uintas and Yellowstone Park! It's possible that I—more than other guys in my shoes—spent more time with my family, but it was not as much as it should have been, definitely.

I gotta tell ya, I have great kids, they are all so smart, and they are so well adjusted and do so well in their lives. The way they treat others and their own children and deal with life is outstanding. I am very proud of each and every one of them. And they were and are fortunate to have their mother. I felt like I only existed in their peripheral life, almost like a phone call as opposed to sitting in the kitchen with any one of them; it was my own fault, and I felt bad because I know what it's like having only one parent, and I hated it! I am glad that I had my mom, but I wish I'd had a dad to do guy things with and to talk about guy things with—it's different. I did have my brothers though, and I was so lucky for that, but I wished all along that I had a dad so I could be his son!

A great guy had once asked me to read a few books (remember?). Later, he taught me about spiritual alignment; it began with God, and then family, then yourself, then your livelihood, and lastly fun. Maybe the order is askew a little, but it places things in a pretty selfless priority. Jack also left me with a challenge, which was about putting the square peg in the square hole. Remember, Tom, the boss and friend who I was going down to HQ (Headquarters) to work for, he had a boss named Bill; he was the maintenance coordinator. In the government, everybody has bosses (lots of them); it's like a giant ringworm, and you couldn't tell where they started or where they left off. It's almost as if for expediency's sake, if one boss breaks, you have a ready replacement; it speeds up the replacement process somehow! What I didn't realize was just how many bosses our office would go through during the thirteen years I was employed there, and more importantly, why.

Anyway, I had to meet Bill (the coordinator) before I took the job; I realized this was *not* an interview, even if it should have been.

During our talk, Bill told me that in time I could really make this place into a home, which seconded something Tom had told me earlier; that with my energy, within a few years, I would be have things whipped into shape! I asked Bill two questions: one, what was the state of the preventative maintenance program? And two, how many open work orders did the electrical shop have? All of which I would be sharing with my staff of one other electrician. Bill stated that we didn't have a preventative maintenance (PM) program for the army side of the house set up yet. Note: the Air Guard had set theirs up fifteen years prior, and we usually got our directives from Headquarters—this place! I was blown away with this answer and told him I really shouldn't take this job. It means that with the age of the equipment, which by the way was located statewide and within some sixty facilities, many of which were fifty years or older, it had to be in a terrible state of repair. Keeping this thought to myself, I knew the army mission and morale was suffering as a result. Bill then told me that we had just two outstanding work orders for the electrical shop. I cut him off asking if it was just for this place or for the whole state; the whole state, he said! Boss, I semi mused, I have never been in this building before, and you don't have to hire me for being so forward, but I can go down any hall in this one place and find something that needs work. I was more correct than I could have imagined! Bill looked frustrated and sorry, but replied, saying, "We really need you!"

My first week at National Guard Headquarters was all I needed to experience to have reason enough to quit right then and there, but I was determined to stick it out and make a difference—the cost and the gain of this decision would be realized over a very long span of time much later. My new boss really did need the help he said he did, but it was to a level that was difficult to accept.

My job as the new lead electrician (as I saw it) was to bring all the electrical systems back to a reasonable, reliable state of repair statewide. This task by itself was a huge undertaking considering how many facilities there were scattered to and fro! Then there was the fact that I was just one of the two electricians the Army National Guard employed with the other one being very dysfunctional, as were several other crew members employed there for maintenance work. I had been tasked with telling the other electrician that I was the new lead electrician; he replied with, "That's bull butter" (or something like that). I told him I was sorry, but it's what I was hired to do and would do my best to get along.

Our office needed me to get control of the flow and quality of the work being performed by the rest of the crew as well. The whole operation needed a major facelift!

The general attitude shared by many was this: we are state workers, what we don't get done today will be there tomorrow. Consequently they accomplished very little on a day-to-day basis. I also noticed a cautious quietness about them that I am not sure I ever understood completely, but I knew this straightaway: I was an outsider and would remain one for my entire stay at this site. I realized with this job what my friend Wayne *meant* by living off the lamb. It was insulting and embarrassing to be considered a taker, guilty by association, but Wayne was painfully correct! However, I was determined for the good of the military and the taxpayers to bite a chunk out of that phrase and change things for the better in my little part of this mutton-faced world. I figured that maybe my attitude and example would rub off on a few of my coworkers too…

In the military, I was taught to focus on supporting the mission, and I always tried my best and worked hard to this end and, along the way, took plenty a flack because of my work ethics, but I don't care! I have to be able to face myself every day and the folks I serve, plus I have always been pretty aware of what I could and couldn't accomplish! After all, I was given a second shot at life and was told my time may be very short, even though I think everybody's is. Well, as luck would have it, I quickly became the bad guy because of this: I worked hard and solved problems—some of them were ridiculously simple, and some were even as old as me; but this is what master sergeants do, they set the example, they find answers, and they get it done, period. It certainly didn't help that my office title, DEH, had adopted the moniker of Don't Expect Help by our customers. We were not earning our right to exist; they had every right to call us names!

We as a group were poorly managed and were not supplying our civilian employees with the right and need to know stuff like the following:

1. job descriptions
2. employee handbooks
3. occupational health and safety classes
4. proper tools
5. safety equipment and training
6. routine work orders

Like I said earlier, there was no preventative maintenance program in place, so there was no routine PM work orders getting into the hands of the employees either! How things worked was when someone would call our maintenance office enough times and complain about some piece of equipment not working, someone would go repair it or, more often than not, replace it. (I can't begin to tell you how expensive this process becomes.) So unless the equipment was dead critical, it might not get fixed till years later since our budget always seemed to be tight—I wonder why. After communicating my concerns with Bill (our coordinator), he agreed to let me help with some of the managerial issues our office struggled with, but he was retiring soon. So to help keep us moving forward, I borrowed the Air Guard's PM (preventative maintenance) program and began inserting it into our operational program. When the new maintenance coordinator, Robert (a retired army captain), came on board, he really helped integrate this process into our system. We obtained a Real Property form from the state facilities office and got some of the crew members to pitch in and identify electrical and mechanical equipment (i.e., air handlers, electrical panels, water heaters, and such)! Literally anything that we were assigned to maintain was identified using the real property form, which in turn listed the important and critical information. Then the equipment was assigned a number and location complete with a weatherproof ID tag fixed to the gear. All this information was then painstakingly entered into the PM program back at the office by Robert and his secretary. Afterward, the computer would then kick out a PM work order on a predetermined maintenance schedule, usually every six months or annually etc. Once you have this program in place, it really helps you schedule your man-hours, balance your budget, and keeps your equipment functioning much longer than it had been in the past, and it saves gross amounts of money. But why should we care, it's not our money, it's taxpayers' money! This last statement came from several individuals on our staff indicating the mind-set.

The cool part is that once you have the PM program running, everything is on a routine schedule; the equipment becomes easy to maintain, and your life becomes less stressful and beats constantly putting out fires—but again, the key is to use the program. Subsequently, we started producing scheduled PM work orders along with integrating small construction projects into our day-to-day operations. The bigger projects were now beginning to be handled by contractors as time and money and need dictated giving me some needed breathing room. We also began managing our grounds and custodial crews better, and those areas were taking on a new look. Before, the headquarters building overall was substandard in regards to cleanliness; it was filthy! After all, don't you agree that a military headquarters ought to look spic-and-span? Again, a management issue!

Old habits, like opinions, do die hard and can be a tough nut to crack.

One time shortly after I took over as lead sparky, I and the other electrician (a master electrician) were responding to a complaint from our ground safety office. Apparently the eyewash alarm in the helicopter battery room at our aviation facility wasn't working. At the scene, I climbed up an eight-foot folding ladder to gain access (through the drop ceiling panels) to the flow switch located in a waterline while my partner—looking bored, sat in a chair! As I approached the flow switch, which activates the eyewash horn, I noticed that it had strike marks on its brass body, which can render the switch useless; just as important, the switch had been installed incorrectly.

Flow switches have a paddle that when installed in a waterline and is moved by flowing water, a set of contacts are closed, causing a light and a very loud horn to activate, indicating someone is in trouble and needs help! The flow switch has an orienting arrow on its body showing the direction of the paddle relative to the flow of water; it is impossible to get it wrong unless you want to! I mentioned to my partner, who was the original installer, that this switch was installed sideways and couldn't work, besides the fact that it had been hammered on. His reply was, "You don't know how things work around here. This is how we create time to go to the movies in the afternoon." I couldn't believe that he was serious. I was stunned. Then he said, "Whether you get that thing working or not, if anyone asks, it was working when we left!" I was used to doing the job right then walking my customer through the finished product, making sure I had taken care of their needs.

I told this "hoser" to forget the ground safety office, what about the person that just got acid in their face, or in their eyes, they will need that alarm to get immediate help! My good partner responded by asking, "Do you really think that'll happen?" I told this yahoo that I have never been very good at predicting accidents. I continued with, "We are licensed electricians and are legally responsible for our actions, and we are human beings just like our customers who trust us to do the right things when we maintain or install their equipment—they should not have to worry about what we did or didn't do." When I finished my rant, my partner stated that he didn't care what he did, who he did it to, or if he did anything at all! Man, I was used to working with folks who were honest, fair, and cared about their fellowman. Yeah, I realize that life isn't perfect but.

I informed this loser that I was going to help him with the "anything at all" part of his statement at the very least. Then I told him that as a master sergeant/electrician, I was responsible to protect my military brothers and would be turning this offense over to the IG (inspector general) Office and would report the incident to our office as well. The punishment he received for his blatant acts were for me to simply assign him work at Headquarters, where he could be watched more closely, which I did under protest. Along the way, this individual and several of his cronies threatened to burn my house down and kill my family including me unless I stopped working so hard; they cried saying it was making them look bad. I told them it wasn't my intention to make anyone look bad, I would rather make them look good; they didn't care and continued with the threats!

I stood my ground by saying, "It really is too bad."

"What do you mean?" one of them asked.

I told them that they knew I was in the military, but did they ever stop to think about what my qualifications and assignments might be? I then told them that I was an Air Force sniper (I lied), and now I may be coming for them, and they couldn't do one thing to stop it. While I talked, I was holding up one forefinger (it's the one-shot, one-kill sign), and one of them knew what that meant—I'm not sure I did entirely, but my acting diffused them a bit. After the confrontation was over, they told everyone I was crazy, but the ruse scared them off, and I could focus on my job again. However, when someone makes a threat, they may intend on carrying it out! So I let my leadership know about the incident and my concerns; they shrugged it off saying they don't really mean it—you don't worry, just keep moving forward. I can't say that I felt very warm and fuzzy over that gesture of confidence. I had already figured these truants would sabotage some of my efforts, but I am a military guy, and this is what we train for, so I adapted. As I made it through my first year on this new job, I found that the road crew / electrical truck had never been serviced since the day it rolled off the lot, which was literally ninety thousand miles ago. Man, I'll tell ya, I hate to walk, and especially because my vehicle unnecessarily broke down. There is no reason to not maintain your ride, considering the fact that we travel hundreds of miles every month and that the state of Utah has a fleet maintenance program that is very easy and accommodating to use. So I worked with the new coordinator, Robert, making sure that we were taking care of our fleet. It seemed that everything under our roof had been left to question, and up to this point, there had not been any accountability. But now the pressure seemed to be mostly on the new maintenance coordinator to fix it or get the axe. The stress on this position was enormous considering all that was on the plate, including the little wars within the department, the infighting, deceit, stealing, slow or no pace, the excuses—it was awful, a nightmarish prison to one degree or another; yet there were plenty of reasons to do good as much as possible, and despite the pitfalls, somebody needed us. After a few months, we were slowly and painfully beginning to resemble a place of business. We were taking on an image of a halfway-interested maintenance operation, and our customers seemed to be having a new faith in us too; it was something that you could feel, they even smiled and were friendly when we showed up. Around this time, I acquired an apprentice, Todd, who had just returned from basic training and technical school for the Army National Guard; he was a private by rank! When we first met, I didn't think Todd was too sure how to take me. He had worked for our office before he went into the army and was, I guess, used to less-disciplined folks. I asked him very brashly, what makes a man crazy enough to become an electrician? That comment seemed to make him stutter, but he replied, "It's interesting…I guess." *Hmm, we'll see*, I thought!

As Todd and I got busy, in a little less than a year, we had repaired and relit hundreds of outside lights at our facilities all over the state. Routinely, we were told that they had been out for over five years! I've heard that line many times before; it usually is an exaggeration with a motive to get things done right now (including things not on our agenda). However, in the case of these facilities, I had every reason to believe them. I also believe that our power bill went up noticeably too!

Not all of my army customers were very receptive of the service they were now getting, but Robert (our coordinator) told the crew: if we work hard and are honest, we will win their confidence. I can say that eventually I was beginning to feel accepted by these folks who were the same ones that our office neglected for so long and even endangered. In fairly short order, given the scope of what lay ahead of me with my new Army National Guard challenges, I realized that I was going to have to shut down the other part of my life, which included climbing, snowboarding, etc., at least for a while; unfortunately, this decision on my part would include my family to a certain degree too! In my mind, to directly affect the mission in a positive way, I had to treat my job like a real-world deployment, which meant to hit the ground running with an elevated OP tempo for maintenance, while receiving ground support for big projects from contractors—if possible. Given the situation, with the small crew and a big workload spread geographically far and wide, I had to be very resourceful to succeed. I gained permission to use contractors as needed, which worked out really well; I was free to work in Northern Utah while a contractor was helping me in Southern Utah, or even locally at Headquarters. Unfortunately I couldn't be around much to check quality control and had to rely on the ethics of the company. To give them an incentive to maintain high standards of execution, I told every contractor this: "I really need your help, so if you do good work and I don't have to fix your mistakes, you will get more work. If you do bad work, you will never work for the state or the military again."

Some companies didn't like this statement too much, but I really didn't have time to repair their mistakes. I can tell you that the way the government does business typically is if a contractor doesn't complete their work or if they do it wrong, the maintenance worker is oftentimes forced to correct the problem. I have personally been ordered to eradicate dozens of these issues, grumbling all along the way that we paid that company to do the job. Maybe we should withhold their pay till we get the product owed us or take legal action against them. The $200 hammer idea is for real and is total BS; it's our money being abused, if not stolen! Any president of the United States who is trying to fix government issues, like funding abuse, has a huge undertaking on the plate. Also, the military never stops evolving! So despite the help from the contractors, I seemed to be getting busier and busier like a truck rolling down a steep hill without brakes. Todd, my apprentice, was very smart and very helpful, but that doesn't mean we had a seamless experience together. One time some troops at a vehicle maintenance shop in central Utah were washing a big dump truck out in their parking lot. I noticed that they had the pressure washer plugged into an extension cord leading to an outdoor receptacle that wasn't a GFCI (ground fault circuit interrupter); you know, they are the outlets with the little red and blue buttons that will save your life when your hair dryer falls or is thrown into the bathtub you are lounging in. Everything was wet—the cords, the cord junctions, and the workers; it was a recipe for disaster. I told them to hold on for a minute. I then had Todd take a look inside the shop at the outlet that was feeding the outside one; he said it was a GFCI. I asked him if that meant the outside plug was protected by the inside GFCI; he said yes. So I had him push the test button that disables the plug; the outside unit did not shut off. Our troops easily could have been killed by this shock hazard! I showed Todd how to easily and quickly repair the safety issue; we then let the troops continue safely washing the big truck. While we were there, we also realized that the eyewash alarm had been wired wrong too. All these safety issues had been created by the electrician relegated to working at Headquarters. I was getting a little tired of the dangerous and stupid things we kept coming across and the fact that our people were unknowingly exposed to these hazards. Todd finally told me that he realized that I had come down to the army side of the house with the intention of correcting all the problems, but then informed me that I would fail—I blew a gasket. I loudly and firmly called the private to attention (which drew a bit of attention from the troops) and informed him that I did not come here to fail, that these guys need us, and I was gonna move forward with or without him; he teamed right up saying he would support me all he could. I really appreciated Todd, and as we grew together, he confided that he had been forced into this job and was being forced to be an electrician. As a military man, his specialty was electronics, so I asked him how well he liked that aspect of the business; he said he loved it and dreamed of running his military shop one day! So after a bit more talking, Todd left his job with me to go to college and become even more proficient and effective for his military shop and these days is in the upper level of that shops management; he became his own man, and I am very proud of him for all he has accomplished.

A few months later, a young man named Jimmy replaced Todd as my helper. Jimmy's personality seemed to be the exact opposite of Todd's; he was quiet and unsure of himself, but as we worked together, he was becoming a very handy apprentice and, like Todd, was also very smart. Both of these kids had told me early on that they didn't believe they would be alive past their thirties; this idea bothered me, and I wondered why their future seemed so bleak. I know I was walking on eggshells myself in some ways, even though this job had buried me with work causing me to forget most of my issues most of the time, or so I thought. But still you gotta try, don't ya? I made sure that I routinely encouraged them to plan and to look ahead way down the road of life, to work on educations, and to realize their individual value and importance to the community, their families, and to themselves and the creator. Maybe it was just a part of them needing some time to mature, but it really did concern me. I also made sure that we had fun on road trips. We would do our job and a little more; then we would eat well and go for short drives, hike, or take out a telescope and look at cool things like Jupiter, Saturn, our moon, or any number of things the sky has to offer; these kids really seemed to eat it up! Maybe in this chaotic world, they just needed someone to care.

The first time Todd, a man of science, saw Saturn and its rings, he flipped saying, "It really does have rings, cool!" On that same trip, Todd announced that he was an atheist. I replied, "Oh, oh, you have been to Missouri to basic training, and that's it for *your* 'out of Utah' excursions, and now you are a twenty-two-year-old atheist!" I told him that he sounded like a recording, like he was repeating something he had heard somewhere or from someone else. I said to him, "Todd, you are a man of science, and science has brought about many great things, and we have made incredible discoveries and created marvelous things over the last several hundred years and especially in this last century." Nodding, he said he couldn't agree more! I then asked him if the dinosaurs were not facts long before we stumbled across them and called them facts while patting ourselves on the back. That Saturn has rings though most people will never see them for themselves, but will say they are there because they saw them in a picture. Also, there are millions of galaxies out there beyond our own, and some of them are just like ours, but is it safe to assume we are not the only game in town? We take so much for granted. The sun comes up on time every day, did we invent that? Mars ain't that far away, and given all our smarts, we have not been there yet (except with rovers), and it will be a big deal to do it in our lifetime, let alone keep our astronauts alive long enough to get there then bring them back home. We are not that great; our opinions of ourselves are always limited to the next electrical storm or hurricane that we can't handle. I told Todd that I really have never liked blindly following crowds; we will get to Mars in time, but it will be done by thinking out of the box and by people who think for themselves.

Jimmy (Todd's replacement) told me one day that he was stupid and couldn't learn. "Really?" I questioned. His mother had told him this artifact and that he should be happy with mundane jobs that didn't require too much thinking—now I had another project. Amazingly for such a stupid kid, Jimmy picked up the basics of the electrical trade pretty fast! I mentioned this to his mother who worked in the warehouse at our Headquarters building; she claimed that he just couldn't learn anything, that he was just too slow and stupid. On one of our many assignments together, Jimmy and I rolled to an armory in central Utah to install some circuits in an office. While we were working, the motor sergeant (Boss) from the FMS (or field maintenance shop; the new army acronym for automotive garages and was probably changed to confuse the Taliban) next door stopped by. The sergeant asked me if I was the new electrician who was helping everyone. I said I was, or at least I hope I was! He then asked me if I would come by his shop and see him before we headed back to Salt Lake City; I said sure. After Jimmy and I finished up our task, we went over to visit with the sergeant at the FMS! Immediately the sergeant asked us if we had ever seen someone killed by electricity; I said I hadn't but had witnessed some of the carnage it can create. He then told me and Jimmy that the other electrician (the Hoser at Headquarters) was the one who wired their pressure washer, including its onboard GFCI and must have done it wrong! The sergeant then told us that recently, one of his subordinates (a private like Todd) was out washing one of their big trucks called a deuce and a half when something went wrong. Apparently, there was a ground fault that, for whatever reason, passed through the kid washing the truck, causing him to go into fibrillation; he died—CPR, EMTs, and defibrillators made no difference. A family lost their son, and the army lost a man. I asked the sergeant if there had been an investigation; he said that the army had determined that it was equipment failure. I didn't believe it! The sergeant then told me that this same electrician had rewired the GFCI outlets in his shop, and they don't work like they should, and they were afraid to use them; they didn't want to get hurt, and our office hadn't seemed inclined to help. The next question was, would I help? Wow, what a question. Of course, I will help, it's my duty! Jimmy and I corrected these issues later that week; it wasn't dangerous, just wired in such a way as to render the outlets powerless when you plugged in a device. In fact, there was so much to correct that I relied on Jimmy to work on his own, to ask questions as needed so we could have our four hands busy. That kid proved the world wrong that day, breezing through all his tasks! I made sure he understood what he had accomplished and that I was so thankful to have him around and to be able to work at that level! I told him, "You never again listen to anybody that says you are stupid." This job proved that idea wrong and closed that door for good!

Back at Headquarters, I counseled with my leaders, showing them the legal pages in the National Electrical Code book. We talked about malpractice and liability and that I had found five other pressure washers that our other electrician had wired that were dangerous. We had to put out a warning to the owners/operators of this equipment to not use it until I could get to them and make the needed repair. The wiring of this equipment was so wrong that it was apparent that no educated licensed electrician could have done it; interestingly, our man in question carried a master electrician's license. I implored the leaders that "we know about him, which makes us responsible"; they weren't too sure about that comment. I told them that at the very least, they needed to let this individual go, or I would take legal action against the National Guard; so this individual eventually was asked to retire early—big damn deal!

Time just seemed to be clicking away, and I seemed to be correcting more and more crazy stupid things in the electrical and HVAC career fields particularly. I found myself traveling alone constantly performing maintenance and doing small- to medium-sized installations too. Some of these jobs were fun, but I had to clip along pretty fast, so the enjoyment level was pretty low. There was so much to do that sometimes I would ask the facility manager (a sergeant) if I could stay at the site and work till I got too tired, sleep on a cot, then get up and get going again. Often this scenario was forced because I had to finish a project today in Northern Utah, then head the next day, hundreds of miles in the opposite direction—sometimes just to meet a contractor. Many of these days became endurance challenges as well as technical wrestling matches.

Another cool thing these facility sergeants did for me was to allow me to store parts for upcoming projects at their site; I even kept the basic maintenance things around too. These guys would help by changing lightbulbs, switch and receptacle cover plates, repairing what they could between my visits; they understood my dilemma. I really appreciated how much the troops helped me help them!

One thing I noticed with my maintenance leaders was that they didn't like to finish jobs! They also believed that we had lots of work to do, and they were right. However, where they were wrong was their thinking that if a worker showed up and put in some time on a project, then this would appease the wild beast; then the worker would be sent to appease more wild beasts. Finally after a year or two, the ten-foot-long and eight-foot-high wall was finally built, painted, and done; and the first wild beast was happy, that is, if it was still the same wild beast.

On the flip side, we tended to cater to the whims of *some* officers at Headquarters (a.k.a. the Flagpole), often to the nth degree! One time, two colonels in the front hall realized that one of their offices was two feet wider than the other colonel's office and demanded that our department make the offices equal in dimension. Our boss at the time, Tom, agreed to do the work while I blew another gasket—what a bunch of jokers, the cost and babyishness of it all. But Tom knew how to feed the rat, saying these are the people whom we take care of first! Man, I am an on-the-ground, tool-slingin' master sergeant and electrician who does not care a rat's ass about such things. Unfortunately in this place, it was the standard; some of these officers had turned themselves into Pharaohs—self-important ones—but in the end forgot what their duty was and who they represented and served. On another occasion, I was headed to one of our outside armories a fair distance away to respond to a gas leak but was forced against my will to install an electrical circuit for a TV first to appease some colonel in the front hall. How does that appeal to you! Did I mention that we had a new coordinator? It was the old HVAC foreman, Tom, the same one who had helped bring me down to Headquarters. Tom had replaced Robert, who had lasted a little over a year and was moved along because he had not been present at a small flood in the northwest corner of the front hall—a lie had been told; it said he had not responded in any way to the problem and just let the cards (so to speak) fall where they may. The truth is, he remained in constant contact with the scenario responders and was not needed at the sight, and everything was handled and overcome in good time. Some leaders don't seem to want to hear the whole story; in this case, however, they did hear the whole story and went with the lie anyway. Robert just faded into the background—how convenient!

I heard an idea one time that went something like this: *They*, referring to "leaders," always believe the first thing they hear—whether it's the truth or not; it makes sense to me more and more.

There was a time when our office was limited to buying parts at very specific contracted outlets in the Salt Lake Valley only—supposedly! I was on a road trip assignment alone one time and needed to make some repairs in Fillmore and Cedar City! While in Cedar City, I had to call SMD (the sister component to my office DEH) to see if they could get someone in our office to answer the phone; I needed to get permission to buy several parts to complete the job. Usually, if you were on the road, it paid to be very self-sufficient and plan way ahead 'cause you really were on your own, if you get my drift. The SMD's office manager took my call asking what was up. I told him that I was really struggling being limited to buying parts in Salt Lake Valley only, that I couldn't always know what I may need without a recon trip! As it turns out, this guy I was on the phone with was the actual say-so guy when it came to where and with whom and how we made purchases. This manager, Barry, set me up with an account at a supply house right there in Cedar City, saving tons of gas, time, and trouble. Eventually we did this all over the state making it very easy to function and get our work done in a more timely fashion; his help smoothed out one of our bigger hurdles. I got my parts, took care of work orders, and the taxpayer too. Looking back, up to the day I retired, I never believed that my maintenance office (now known as CFMO) ever as a whole wanted to succeed; maybe it was by design or not, but our funding was always used as one excuse…hmmm. I believe in at least trying!

Consistently over the thirteen years I was there, including eight coordinator changes and five directorate/colonel changes, and notwithstanding different management styles—we as a whole repeated our mistakes and just continued to look stupid. However, there was one exception; this coordinator wanted to make a difference, but he was eliminated within his first year as our seventh coordinator. One coordinator, Steve, who came along just before number seven lasted all of one week! He walked up to me one day and said, "You are a master sergeant, right?"

"Yes, sir," I replied.

He said, "I am also a master sergeant (a marine type), this place can't be for real."

I never saw him again.

Here is a status quo reality: it is wrong to do it right, to go against the flow or status quo is certain death. This statement was made to me by an electrical engineer concerning an electrical grounding issue; it was based on the premise that none of the other equipment on the roof in question had been grounded properly. So he didn't feel obligated to do his job any differently. Routinely, I would be told that it seemed like I was one of the only ones on our staff who was home, lights on or not!

It's sad to realize that our current president says he wants to fix our budget when right under his nose is such waste of money and resources—along with no lessons from hindsight learned, along with a casual attitude toward our country's operating funds—are we really that stupid! It's not just his dilemma, it is ours too! We helped create the problem too and kept it alive with our habits. Our leaders live like kings! They indulge in attitude, cars, clothes, furniture, carpet, jets, and fuel—complimenting it with overstaffing and overcompensation while looking you in the face and complaining about the deficit.

Congress (our pals) is always ready to give themselves huge raises, and amazingly it occurs during the worst of times, am I wrong? You people are pathetic; we should impeach you for not leading by example! The government shutdown of 2013 was an example of who we as taxpayers should remove from their positions; then vote in some folks who will do those jobs. What an embarrassment!

The only reason you have a job is not enough people care enough anymore to correct the problem, and some are even afraid to speak up because of possible ramifications—this perversion runs deep. Our leaders need to earn our trust the hard way by affirmative and correct action that works for everyone. Despite all the apathy that seems to pervade just about every government office, I guess we do pretty well; life could be worse, and it is for some!

Okay, off the soapbox. During my employment with the Army National Guard, I got the opportunity to grow exponentially as an electrician. Life is about perspective, and the upside to all the drama was that I had, in a way, an electrician's dream job. I wasn't stuck out in the construction fields doing the same thing day in day out, freezing all winter and roasting all summer, looking for work when business was slow! With my military affiliations, I have never been bored; in fact, I believe that when the world ends, the army will still function for several days after based on inertia alone.

I guess I could have lain back a bit and went with the flow, and nobody would have been the wiser; but there were opportunities to become proficient with rising technology, including the following:

a. harmonic distortion and all its challenges, including general power quality
b. large and small UPS (uninterruptable power supplies), sometimes called battery backups
c. solar farms
d. wind farms
e. IT operations and server room design and installations
f. LED technology
g. power generation
h. power quality—harmonics, power factor, and TVSS (transient voltage surge suppression) systems and related test equipment

There were also all sorts of specialized equipment that I got to design circuits for, then install. The army asked if I could help, so even if it meant going to school so I could, I did! Tom, one of my many coordinators, said that he had a great deal of confidence in me and thought that I should take advantage of the learning opportunity and run with; it ultimately would turn out to be a good thing. Despite the rough and tumble nature of this place, I guess the perspective I maintained was that I would take the advantage of the learning opportunity, but I kept a little perspective open toward the wasteful and stupid and dangerous things that bugged me. I wanted to fix them—which meant I was opening a door to quite a bit of stress on top of everything else that was going on. It was to me a small price to pay to keep the mission going forward, helping out the taxpayer, and trying to forget the ninth of August 1990. However, I really had lost sight of how fast time was flying by and what was happening to me because of my decisions. I thought back to a day when Keith and I were bouldering in Little Cottonwood Canyon; he warned me that if I went to the army, it would steel my life away! It was the next spring after I had indeed took the position with the army that Keith and I met for some bouldering. My pal simply commented, "Let's get into shape for the climbing season." I snapped harshly at him—barking, "I am in shape." He calmly let me vent about how things were going while I moved like a robot over the traverse on what's known as the Standard Overhang Boulder. As I climbed, I thought about Jack trying to teach me about keeping the square peg in the square hole. Exhaling hard and frustrated, right then I knew my climbing days were over; they evaporated into the air and were carried away on the winds of my decisions and change. Keith had come into my life right when I needed him to; he understood suffering and pain and goals. He had fallen 150 feet to the deck off an ice climb in Banff. After regaining consciousness, he gave himself a damage scan (he was dying) then snapped his partner back to reality so he could help! Keith (balancing on his ice axes) hiked out several miles before he collapsed pretty near the trailhead! Keith had a broken leg, and several broken ribs stabbed through his lungs. He taught me all about patience and toughness. I never got to climb with Keith again.

Along the way—or, more accurately, with my war against the army and their way of doing things—my oldest brother tried his best to teach me some lessons. He would hand me a piece of paper and a pencil; then he would have me draw a one-inch circle somewhere in the middle of the page; then he would have me draw a four-inch circle around the smaller one. He knew that his little brother was in a battle with his surroundings at work because it was all I thought and complained about, and it was consuming my life! He would tell me that the big circle represented everything associated with my job, and the small circle represented what I actually had influence over. Secretly I disagreed; I thought if I persevered and stayed focused, my circle of influence would be much larger than the small circle—when in reality, I could have cut my circle in half, and I would not sustain much influence there either. I had become a slave to my job and worked harder and harder to prove my worth, satisfy my sense of duty, and maybe even cure some of the ailments associated with the lightning accident. As my other life all but disappeared; I arrived at retirement realizing how wrong I was about many things! I was trying to survive and succeed but had lost my balance in life; it explained why my old habitats (short of the electrical world) didn't seem natural to me anymore. I felt like an unwelcome visitor rather than a piece of the pie. Keith told me one time that I had forgotten how to have fun! My stress from work was skyrocketing, and I was hurting my neck even more with my so-called diligence of accomplishing things in the "right now" mode.

Due to the way the headquarters building evolved with its succession of remodels, the structure contains over two hundred electrical panels with over four thousand lights along with plumbing and HVAC equipment installed overhead and above ceilings and almost impossible to get to! Most often, lots of modular furniture blocked the initial pathway, so I would use my climbing skills to cause the fix to take minutes instead of days. Why were these important design issues never factored into the scope of work? Making some of these repairs caused me to have to get into positions with my body that really worked my damaged spine; it wasn't much easier for my coworkers either. My leaders were always more interested in not wanting to hear about the challenges as opposed to that the issue was handled immediately—more status quo. I was breaking down physically and would get repaired, then just jump back into the war! I told my supervisor Tom one day that I was noticing that I was beginning to wear my emotions on the cuff of my shirt. I was changing, and I am not sure for the better and didn't like it. I was used to planning my work and working my plan, but we were moving at such a pace that I was starting to hope that my instincts caused me to do the right things on all the jobs I was engineering and then installing. I often wondered how come I didn't get a speeding ticket as I flew around the state. Little Jimmy had left our employ to explore a career with an electrical contractor where he could go through an approved apprenticeship program, and the other so-called electrician finally retired, but the Guard didn't hire his replacement for over a year. Finally a great guy and electrician named Clark who had replaced me after I left the Air Guard came to work with me at a time when I was about mentally toast. He helped me tons, and I tried to make his life in this nuthouse as pleasant as possible by assigning him the outlying work sites; we also would work together as needed, which was quite often. We made a great team and accomplished lots of good things for the army folks. By the way, do you know that it's true? Misery loves company!

Some of the company I kept besides Clark was our energy manager Colonel Luke, who was a captain at the time, and Sergeant Duke, who would later become a chief warrant officer and base engineer at Camp Williams. I first met Chief Duke in the late 1990s out at Dugway Proving Grounds in Utah's west desert while supporting the Global Patriot training mission where we worked together to get a command post operational ASAP! Sgt. Duke eventually was transferred to Camp Williams in the valley south of Salt Lake City, but has worked for our department off and on over the years. The three of us have been hooked at the hip for well over a decade; we did our level best to keep the basic maintenance program I brought down from the Air Guard moving forward. These buddies of mine suffered from the same circle of influence ideas that I did. We had great ideas collectively but did not contain the authority to put them or keep them in action—even though most of our agenda were outright operational requirements! With each new coordinator change, including the directorate/colonel changes, our goals fragmented closer and closer to becoming a square wheel; continuity was almost impossible. Then with our second-to-last leadership change, the idea of being organized and consistent became a dead horse. It's so frustrating to see a square wheel start to become round then to just have it become square again. Amazingly our leadership would tell us that we were poorly funded, supposedly only to 50 percent of our requirement, so here we go again. I knew this was a ruse to acquire even more money using (we'll say) excuses. After all, if we are so poor, then how do we explain all the forty-eight-inch TVs and smart screens Clark and I were busy installing, and I mean literally dozens of them. What about all the latest modular furniture we installed when the existing ones were just fine and were maybe even only one or two years old? How about all the LED flush mount–style fixtures we installed in our hallways that cost six times what a fluorescent style cost with a payback that is well beyond the seven-year cycle set by the president of the United States himself for energy conservation? What about the multimillion-dollar windmills at Camp Williams that are rarely operational, which will never give a payback? Ya, we're broke? We can waste all that time and money on that crap, but we can't have a standard maintenance program to follow, have training for our staff, proper tools, introduce a safety program, etc.; the list is so long with the things that should be present that I just want to throw up. I sincerely hope that we are the only ones in the country who operate this way—fat chance! So we increased our managerial office staff, probably with the idea in mind that if you make everything look good on all those government forms, then everything accountable for must be just fine, and/or it gives the leaders someone else to blame. We also made sure we hired people who were less than qualified for many positions too, even contractors! This is the only way to keep the viscous cycle strong and enduring. At one point, a ploy of getting behind was used to try and get more funding from Guard Bureau. The idea was to have the bigger office staff do a report on the maintenance staff that was supposedly shorthanded and show a bunch of outstanding work orders with decaying facilities etc. I wonder why our funds went the way they often did, toward useless things like bigger office staff or to contractors who wouldn't finish a job or do it right. I wonder why we didn't try working first. Remember me saying that the first story told—truth or not—is the one they will believe?

One of our engineers, Rodney, was the project manager over a remodel of the DOL (Directorate of Logistics) department (more acronyms/more confusion) at Headquarters and had hired a local electrical outfit to do that part of the work and seldom involved me. A few days after the job was complete with everyone paid and the DOL folks back in operation, I got a phone call from their chief warrant officer. The chief asked me if I would come down and have a look; one of their portable battery backups for a computer had caught on fire and burned up! So I grabbed my tools and went to see what was up. When I got to the computer, I noticed that the battery backup was an older model, and they were known to have this issue in their old age! Out of curiosity, the chief and I took the unit apart and were stunned to see how damaged the circuit board was. I told the chief that it was high voltage that caused this, yet the outlet that it was plugged into was running nice and steady at 122 volts. So we started looking around a little more and found a surge suppressor—the kind that plugs directly into the wall—just burned to a crisp; the wall space around it was blackened to about the size of a cantaloupe. So I took the outlet apart, noticing wiring flaws that were not the cause but were a potential danger anyway. As I poked around, I eventually wound up back at the electrical panel and transformer that supplied power to this area and found the real problem! The electrician who connected the transformer and panel had created a floating neutral, which can cause wild voltage and current swings, damaging equipment and starting fires—amazingly all it came down to was a two-dollar piece of wire not installed in the transformer. Interestingly, I found out that an apprentice did all the wiring and had run this job without the presence of a licensed journeyman. I alerted our engineer who had the company send the unqualified kid back to fix the problem; after five separate attempts, he kinda got it right. On this fifth try, his boss came to inspect the problem, along with several state facilities representatives (the folks who manage all state buildings) and a respected electrical inspector I had invited. Also present was myself and Chief Duke who worked for our engineer, Rod. I showed everyone present that as it stood, there was dangerous electrical current flowing on all the associated metal surfaces of the equipment (which could be touched by anyone walking down that hallway), which was due to a separate illegal wiring issue this electrician had instigated; then I had the inspector verify my claims! I brought up the fact that it was illegal for apprentices to work alone without supervision and recommended that the electrical company not be paid till the work was done correctly and that the electrician(s) be disciplined. The state facilities people asked me, "What are you, a walking code book?" Our engineer at that point told me, "If you and Duke want it fixed, do it yourself!" After that, he went in and told Captain Luke (our electrical engineer technician) that I am an unqualified electrician and should have me removed from my position and demanded a sit-down with the three of us to prove his case. So when I responded, I stood up to the accusation showing my license to practice that was earned and kept the hard way, then asked the colonel to look at the copy of the inspector's report and pointed out that our engineer didn't want to upset his contractor! I also made the point that if I am so unqualified, why is it then that this same mechanical engineer, who on his own would sorely lose an electrical debate against me, wants me to rewire his house? Man, it got quiet. Later, the engineer showed us by firing Duke from project managing because he supported the demands against the contractor. I got off the hook, but I was some sort of bad guy 'cause I didn't or wouldn't play these games. Nonetheless, the three amigos kept staying after hours working on projects and trying to come up with a plan to fix our broken world; we did at times wind up having some authority to make a change, but it always had to meet the next roadblock up the chain of command where it may be turned down. We lost so much family time, and this didn't include all the deployments we were all going on either. I can't believe I was still field-testing products for numerous outdoor companies too; where did the time and the ability to be objective come from? Our great repair plans at work seemed to fall on deaf ears! All that time and effort—gone, but it was replaced with frustration and stress, and I don't think we ever noticed how little by little, the rat was eating away our lives. I was losing sight of me, of who I was! My goals and dreams along with my real responsibility—my family—were riding away on waves created by that rock I threw into the water (so to speak)!

Recently, Luke (now a colonel) got deployed to Afghanistan! I received an e-mail asking if I could fix a few things at his home that he didn't have time to get to! While talking with his wife about the work, she asked me what retired life was like. Remember, I retired from the military side of the house in 2008 then retired from the civilian side of the house in 2013—great, I said. I told her that everyone who knew me believed that I would show up for work on January 17, 2013, the day after my last day, just out of reflex. I told her that the truth is, I slept in and did not think about the nuthouse even once, and to this day rarely do. Don't forget, I was one of the three amigos, a.k.a. the driven-to-succeed people. I then told her what my second oldest daughter said to me at my military retirement right in front of the general! As I was going to the pulpit to address my clan, Melanie (my second oldest daughter) grabbed me and said, "Dad, do you realize that you missed all of my birthdays that I remember because of the military?" She was thirty years old then. I felt terrible, while the general fidgeted uncomfortably in his seat. My talk to the crowd changed to lecturing about sacrifice and duty as opposed to all the good times we had had along the way. It was a hard pill to swallow, and of course it was just one part of the story. I also told her about how my wife reminded me recently about how lonely she had been the last thirteen years since I had been employed with the Army National Guard. All I could do is say I was sorry as there was no point in defending myself. I knew that my wife had been worried that she would lose me and be left alone for good! Mrs. Luke asked me if I thought her husband will be able to walk away; it has been twenty-six years with him being gone all the time. As she said this, Thanksgiving was only a week away, and he was scheduled to miss the next one as well! I agreed with her that he is very driven to duty and the mission. I told her that I believed that he would be able to leave it all behind for her and their kids, but it wouldn't hurt to start him thinking about it soon. It just seems that we get in such a big hurry anymore that we just blow right through the stop signs, and sometimes we get lucky and sometimes we don't…

To add to the stress during the 2000 decade, as if 9/11 wasn't enough, my mother and father-in-law died while I was deployed on missions!

My mom had incurred five heart attacks four days before my unit was rolling out to a mission; it was the Wednesday after Mother's Day of '04. There was also a mass in her chest, I assume was cancer; she was not expected to live, so I was allowed to stay behind. Two days after she had the heart attacks, a pacemaker was installed in her, and the doctor told me to get back to my unit and go on the deployment because they needed me, saying that my mother would live another twenty years. My brother Jess, who lives in Illinois, called me that Saturday night and asked if he should fly out since he hadn't seen Mom in several years already. I told him what the medics said and told him that I was leaving in the morning on my deployment! My mother passed away the following Tuesday! The military got me home in about three days, but I was crushed and felt terrible for Jesse because he didn't get to see her before she died, and I felt to blame. My father-in-law died of stomach cancer less than a year later while I was deployed. This time I was not allowed to come home and be with my wife and the family to grieve with and comfort them. Throughout the first decade of the 2000s, I lost several friends (military) and two teenage cousins (a sister and a brother) to suicide, several friends to cancer, and two handfuls of friends to climbing accidents due to falls and avalanches. My family had our share of life-changing events all through this period. One of my lost friends was a marine who fought in the Battle of Hamburger Hill.

Another time, a very close friend of mine—along with his terminal illness—just couldn't take life anymore; he set his keys and security badge on his desk, drove home, and died. He was a father and grandfather, a Vietnam hero, and a great friend gone. This humble man was amazing; he was the best carpenter I ever met. He was calm, cool, and an ultraquiet person who used facial expressions maybe more than words, and I loved playing tricks on this guy! One time, several of us Air Guard guys (long before I went to HQ) had a calligrapher help us change the measurements on his twenty-five-foot tape measure for the first several feet in quarter-inch increments (yes, it was elaborate). We then changed this present month's page on his desk calendar with last year's—changing things for him by a day. One afternoon several days later, my friend was at his desk mumbling, saying it just doesn't make sense. I asked him what was wrong. He said that all week long, he had been a day behind and a quarter inch short! My accomplices and I tried to bite our tongues and cheeks; however, our body English helped it make sense despite our vehement denial of any knowledge of, or complicity with, such things. Our buddy knew better—nodding and smiling saying "uh-huh, right!" This quiet gentle giant paid us back by making us wonder and worry when the day of reckoning would come; it never did, but for the wrong reason. Another friend, Sgt. James (our unit supply sergeant) and his family had been a prisoner of war in Wyoming after the attacks on Pearl Harbor! Sgt. James's wife died many times while battling pancreatic cancer, and he struggled right along with her, going to her side until he was threatened about losing his job. Finally she left her husband; then he left us after thirty-eight years of service to the country and never came back.

Think about what I am saying: kindness or brutality both go a long ways in reaching out and can ride the ripple wave forever—you get to choose. My brother Jesse's wife and daughter both died of cancer just two years apart, and in the same time frame, he lost his best friend to an artery defect; then a close uncle and father-in-law. Yet my brother's faith remains and is greater than ever. Recently, he was asked what the most important day of his life was. Jess replied, "The day he was born." Then he finished with, "The next most important day was when he realized why he is here and what he needs to do with his time on earth."

My own body also seemed to be losing the battle little by little to the effects of the lightning strike's related injuries. I was told way back in the days of the burn unit that I would experience nerve issues and would damage tendons, muscles, and ligaments fairly easily. As it turns out, I severely tore my left rotator cuff in the late nineties while running a drill motor. What amazes me is that as the drill bound up, I caught it before it could get out of control, but right then, I heard a faint pop in that shoulder and wondered what it could have been. Soon the pain was excruciating! What is so infuriating is that by that point in time, I had regained and surpassed my strength levels from before the lightning strike, and now was about to go through losing it again. I think in retrospect though this really was the beginning of the end of very technical climbing for me, but the decade of the 2000s would finish me off anyway! The sad part was that this injury was misdiagnosed as a sprain, so months passed by; and despite therapy, lots of pills, and encouragement, my condition worsened. One day while at therapy, the physical therapist (PT) asked me what was going on with my back. I asked why and said that my back seemed to be hurting more than my shoulder lately! The PT then told me that my left latissimus muscle appeared to be mostly gone, atrophied! My upper lat had died on the vine due to an impingement by a cyst that formed in the rotator cuff tear (a bodily process designed by the red gods to protect the tear), permanently killing nine of twelve firing units in the upper part of the muscle! I built strength around the problem, but still, I have a noticeable weakness when I perform certain motions like gripping a bundle of electrical wire and pushing it upward into a conduit; my arm tends to give out after only a few minutes. My neck injuries became very active all throughout the 2000s. I was told by one doctor that my nervous system was busy trying to kill me, mostly thanks to my neck; he believed I could die if appropriate measures weren't taken. I know this much, I had days when I felt like I was gonna die; I had terrible headaches that by the end of the day left me weak and exhausted causing me to want to get to sleep as soon as I could so I could survive the next day. I had a handful of doctor's work on my neck over the years. Typically, I would wind up in physical therapy and was given a prescription of pills then sent on my way; shortly after, I would be struggling again. Surprisingly one of these doctors (who was a decent orthopedic surgeon) thought that I should see a sports chiropractor named Lanny Nielson; this guy worked on all kinds of famous people from country singers to basketball stars. While I was in his waiting room (it was around 2003), I could hear this deep booming voice telling people that he was gonna break them in two and they would be fixed. Looking for a way out, I thought, *Freakin' great, as if I don't have enough problems, now I gotta wrestle Spartacus, and I'm gonna get fixed—broken in two, that is.* When Lanny brought me in, he explained that he had seen all my MRIs and X-rays the orthopedic surgeon had to offer, and then he told me, "Look, I can't repair you, you will someday need surgery." Then he showed me a spinal column model detailing my injuries and their potential decay over time. He then said that he can help me be a little more comfortable for the time being and could help me gain some more left and right range of motion in my neck. Presently, I was only able to turn my head left or right a few degrees before it would lock up. Lanny took good care warming me up for the kill, then gave me a jerk that made something snap, which sounded just like a yardstick breaking over a knee! So I went total theatric, swearing my neck was broke and that I was paralyzed for good! Grinning, Lanny said, "No, you ain't!" and had me try my range of motion. It was great! I could finally—after some twelve to thirteen years—move my head side to side in relative comfort. I was reassured that it was a temporary fix and that when I get stuck again, I'd just show up so he can repair me—he wanted to help. Thank God for Lanny!

Interestingly Lanny invited me on a trek to Mount Everest Base Camp that was guided by two climbers I knew, Adrian and Alan Burgess! The idea was that when they got to the mountain, the climbers would climb and the loungers would lounge. I declined because of all of my foot issues. This was maybe my last chance to touch Everest; I wondered if the guides might have at least let me carry loads up to some of the lower camps—damn!

In 2006 while on the job, I had a very close call, I nearly fell sixteen feet to the concrete in our USPFO (United States Property and Fiscal Office) Warehouse—how about them thar acronyms, woo woo! I was working on a light fixture and had stood on the side rails of a rolling ladder to gain a few more inches.

While looking up and reaching, I temporarily lost consciousness. When I came to, I was standing in balance with both of my arms swirling at the side of my body; my two workmates Ron and Glen below were pleading with me, "Layne, don't fall, please don't fall." Our boss, Tom, told me it was time to find another surgeon who would listen and help me—what luck, I complained! Lanny had moved to Missouri, and the other guys just wanted to keep me in the fitness program at the physical therapy room and give me pills—you doctors listening? For months on end, I struggled just to make it through the workday; then I'd go exercise, then meet with the amigos to fix the world, then I'd go test products, and then get up the next day for some more "at jet-speed fun."

One day my wife, Teri, who is my constant guardian, made a suggestion. She said, "I know that you are getting tired of talking with doctors," but that she had heard about a surgeon at TOSH Intermountain Spine Institute in south Salt Lake, and he was supposed to be very successful with my type of injuries. I thought, righty-o, pally! But to humor my good wife, I went, or I should say we went in to see Dr. Berry during April 2006! I told him my history; then he had me X-rayed and told me that I had some serious issues and ordered an MRI, which gave the greater detail he was after. Later when we met after the MRI, the doctor told me that I was awfully close to living life in a wheelchair, that the damage in my neck was extensive, and I needed surgery as soon as possible to avoid potential paralysis. He told me that my nerves that run along the spine were being pinched in between the damaged vertebrae and was responsible for me passing out, dropping tools, etc. We would need to fuse at least two levels of vertebrae and add a plate that ran roughly half the length of my neck, thus putting me out of commission for a while. The doctor wanted to talk about the new set of limitations I was going to have to adjust to as well! My climbing, snowboarding, and military days were over; I couldn't risk getting my head banged by rock or ice or taking any kind of fall, be it climbing or snowboarding. Carrying backpacks was going to be a challenge too!

The military had an issue with the fusion and the plate and ended my career in March of 2008. I had served for thirty-two and a half years and had been shooting for a forty-year career, which is quite a rare and lucky achievement, but as our medical chief, Sgt. Tomshack, told me, "Layne, we just can't keep up with all your nuts and bolts." I was worried about my civilian job giving me the boot as well, but the doctor thought that it would be up to me and how well I managed that scenario to see if I stayed or had to leave. Again I felt trapped and afraid. I knew that I had to apply the formula of staying in shape, but also keeping my mouth shut if it hurt. I had to protect my livelihood for the good of me and my family. I thought again about what my bad decisions eighteen years ago had brought me; now I had to tough it out to the end of the war, which by the way was far from over.

The plate that was installed really limited what I could do; my electrical career was now in jeopardy. I was told that looking up and reaching above my shoulders was two of the things I was supposed to avoid, including impacts to the head and neck. Great, I'm an electrician, this is my life, and this is the world we operate in; and for me, it is ending—no more contact-type activities. I guess my doctor didn't realize just how vulnerable I was even on the job, at least with the army. Like I said before, I decided to keep up a high level of fitness and keep things status quo at work; no complaining (too much) about my deficiencies—nobody was gonna cut me any slack, and they didn't. Even after I was sent back to work on light duty, my colonel told me that if I wasn't 100 percent, to go home and come back when I was. He's the broken record, and I'm left singing the "Seems like I'd Heard That Line Before" song. Sometimes, however, abuse can make you do things—like react! My colonel's comment made me mad, so I worked harder than I should have to teach that yahoo a lesson; meanwhile, my neck began its slow but sure glacier-like destruction of the area above the plate. Apparently when you have a plate like mine installed, the sections not supported by it take more load than they should and begin degrading, eventually repeating the original symptoms my lower neck exhibited. On top of all of this, my old Guard unit was falling apart too! It had all started with the Iraqi war—our leaders assigned people into critical positions who, when the heat was on, failed to do their assignments. Our unit left Utah sooner than they should have and didn't bother to take their weapons to the war—well, rocks don't win in a gunfight. Our troops were also being required to study career progression courses in a war zone (like the war was some sort of video game) and if not completed could end a person's career; with this thinking, we were sinking into trouble. We had high-level sergeants degrading the commander in front of the troops. Back at the home station, these same individuals stole military man-days to gain bigger retirements. All these things conspired to cause our unit to lose its military bearing, including four commanders in very close succession. Speaking of which, I think military bearing is becoming lost across the board, military-wide. We need, as a country, to become more disciplined and selfless. Everything is about me, me, me and is a bit too much cyber related—just pop in the batteries, set a program, and go—we don't think anymore! We need to become more survival minded again. We will have to defend our land again and again! People need to earn the right to have the jobs they have and the property they have; somebody has to pay for it—it should be the owner!

Predictably, my old unit lost its way and now must rekindle the correct spirit to survive a potential unit closure! It's not uncommon and quite routine that commanders, even generals and presidents, don't know the truth of their folds, about what is actually happening on their watch! It could be that there is pressure from above forcing certain priorities even to the point that we must waste time and resources—sounds very screwed up, but it is also all too commonplace. Again, the top leaders in some organizations often let their middlemen tell them how things are. So the facts of the matters can become quite polluted because some of these lower echelons have their own priorities and challenges—to be fair, I will retract to, at times! Nevertheless, from my experience spanning thirty-eight years of service overall, the National Guard, a major portion of the time, does themselves proud at the NCO (noncommissioned officer) and officer level. The whole team does make good things happen.

As you may recall, back around 2003 or so, we replaced the troublesome electrician with Clark; he was assigned to the road crew with one of our new HVAC specialists, Joel. It was a good team up because they had worked successfully together before at Geneva Steel for many years. In the spring of 2008, right after I was medically retired from the Air National Guard, Clark and I traded assignments; now Joel and I traveled the state together. My orders were to bring the St. George National Guard site up to a higher standard; and since this facility resides in a harsh environment, it was time to spend some money on this resource and make it like new again. Being on the road crew can be hard as you travel quite a bit, so you miss being around family, friends, and personal interests when you are out of town routinely. But Clark needed a break and would be working with the fourth new coordinator, Rick. We had also recently hired a new foreman named Braxton, who we later learned was a shoo-in and seemed to have lots of issues when dealing with people. This guy had just finished his HVAC apprenticeship and was jumping ship to be a manager and turned out to be grossly unqualified for his new position. The National Guard—true to their nature—didn't use the probationary period to correct his actions or replace the brute; they just let this individual put the leather to workers and customers alike, which was too bad! Without his presence, the stress at Headquarters on its own usually shot through the roof and rang the mission bell; now the bell went into orbit. With this brute around, the crew that had been pretty reliable walked on eggshells as they were always being threatened to be eliminated. Our customers, the army, took quite a bit of abuse from this brute too, but they fought back and forth against this rude egg filing complaint after complaint with HRO. Around this same time, I had tested a pack (the Alpinisto) for Wayne. During the test, my son Eric had taken some photos of me with this pack; I couldn't believe how old and hammered I appeared. I should have noticed the effects of stress sooner, but more was coming, all compliments of the army and the foreman and his regime.

Lately I hadn't been feeling well, so I went to my internal medicine doctor to get some help; he noticed that I didn't look healthy and asked what was going on. After I was done venting, he shook his head and recommended a new job instead of pills. The doctor warned me that my serotonin was through the roof and I was headed for big trouble like a stroke and such. All I could do was breathe, work hard, hope, and pray.

Within several years of receiving our assignment (2010), Joel and I had the road sites in really good shape per the marching orders from the colonel; our customers were relatively happy, and we were too. I had also decided because of logistics to quit field-testing; I just couldn't give it the true attention I needed to. I was distressed over this decision, but it was the right thing to do at the time. I guess depression was closing in on me all along, and it was strangling me pretty good now; being with Joel helped quite a bit though, he was smart and fun to be around, and I learned quite a bit from the old fart too! Around this time, the present coordinator, Rick, was moved along for various reasons, so our foreman assumed the coordinator position for about five months, immediately ramping up his leatherwork correspondingly. Finally, the Guard brought in another individual named Kent to take the helm of coordinator. Supposedly he had spent some time in the service, and I was supposed to be impressed, but I figured we may have something in common. After our staff's first meeting with Kent, I knew that something wicked had entered our realm and teamed up with the brute. I feared we were all going out the door dead or alive as Kent had an idea and an agenda and didn't care about protocol, honesty, or what the Guard thought either. His idea was to eliminate the crew as it was and hire several of his friends; then he would fudge on the paperwork, making them all leads—and were not there to work, but would manage contractors to do the maintenance and construction jobs. I warned him that our budget would not handle such an expense and would ultimately cost him his job; he seemed unimpressed with my comments. Interestingly, he had already attempted this process with a county organization and was moved along because of it; Joel and I had learned this from some of his former employees who we routinely ran into at parts houses. Two guys in particular whom we saw routinely told us that they knew who our new boss was and named him, then gave us a play-by-play routine that would unfold shortly. They told us that Joel and I would be separated, and then we would get bad performance reviews; also, the fully ready-to-work trucks we currently were assigned would become basic pickups with no bins. Finally, he would start finding ways to get rid of us one by one and, if he could, as a bonus, cost us our retirements. It was hard to believe such rumors, but those guys had lived in that world for a while, and shortly it all began to come to pass. We learned that this individual had previously worked for the state government folks, who apparently let him go for "less than desirable" performance. One time, I asked this phony why he didn't stay in the active duty service and make it a career; after all, he was an aircraft technician with nearly a decade of service. His answer was a red flag to me; he claimed that there was no upward mobility, nowhere to go in his shop, and they were getting rid of him!

In ten years' time, the service has spent several million dollars training you, they get rid of folks at nineteen and twenty years, not ten; they need to get their pound of flesh. Being a master sergeant with a bit of background, to me, this meant there was a problem. I had sensed the first time that I met him that he was going to get rid of us all; he didn't even try to remember our names, nor did he ask about our families either. It's weird though, I don't know how I sensed the foreboding because the meeting seemed friendly enough. I guess it was when I looked into his eyes and saw the real stuff he was made of; I also saw what he meant to do. In twelve months' time, he managed to eliminate twelve of us employees one way or the other—including secretaries, groundskeepers, one electrician, and HVAC people. In the meantime, he had hired four personal friends to start fulfilling his plan, who buy the way, made sure that the several original crew members left knew that we were history. The work trucks disappeared, and we acquired a few standard pickups to work out of. They had no bins, no ladder racks, just an open bed, perfect for a drive in the country on a sunny day. The threat of being eliminated created such stress throughout the crew during the cleansing period that everyone who was a target was becoming ill. Our military vehicle maintenance shop, known as CSMS (combined support maintenance shop), had it out with these leaders and HRO and threatened legal action for the way they were being treated. Other groups complained to our colonel and our general as well because the rude substandard treatment they were receiving was simply unacceptable; it was creating an unfriendly and uncomfortable environment throughout the entire workplace. We prisoners had always said that every time we think we've hit the bottom line, some new manager would come along and take us lower! Now we all wondered if this was not actually the bottom of the pit and hoped that there wasn't a place lower than where we were! In our case, three months into the cleansing process, the original crew met with HRO to plead our case. Soon afterward, we and HRO met with our colonel and the offending supervisors who promised they would play nice, but they chipped away anyhow like rats until there were finally just three of us left. Besides abusing their position, these new civilian leaders abused our budget too by driving to an armory in the southwestern part of Utah to accomplish a so-called task; then, they would stay through the weekend on the taxpayer. As it turns out, they had friends and family in that area, how convenient. In all the trips over all the years that I had to roll to that location, we never had an issue that kept us past our last workday of the week, never! This situation, which involved food, fuel, and hotels and actual tasks accomplished, was being monitored by somebody important; they were paying attention, and it would be exposed, but not until the next military change of command for the directorate of our department!

Around August of this wonderful year 2010, my son and I went on a backpacking trip to a place known as Brown Duck Lake, which sits in a pretty basin at roughly ten thousand feet in Utah's high Uintas primitive area. While on the trip, I contracted HAPE, better known as high-altitude pulmonary edema. It didn't make any sense that I would get an altitude-related illness because I had been hiking all over the Uintas in my spare time and was completely acclimatized. On the hike into the lake, Eric had begun feeling sick. Shortly after, we arrived at camp; we ate dinner while enjoying our campfire, but Eric was still under the weather and went to bed early, but I felt great. This trip was just what the doctor ordered in lieu of the stress at my workplace. I knew I would be able to find some balance out in the hills, even if I was more at odds with life now than I had ever been! Given the events of the last ten years, I was ready to jump off the fence anytime; it was hard facing the daily music, but deep down, I still believed things would improve. Change is always a constant in our universe—it was this hope that kept me hanging in there.

Anyway, the next day when Eric and I got up, he looked all rejuvenated, so we went fishing at Kidney Lake a few miles away and got skunked; it was fun anyhow! While hanging out at Kidney Lake, we had a grand lunch of salami and cheese then took a nap. As the day progressed, we headed back to camp; then Eric noticed that I was just kind of dragging along and asked if I was all right. I responded, "Yeah, I didn't sleep real good last night, but I'll be fine." After we got back to camp, we cooked dinner and then fished at Brown Duck for a bit then lounged around the campfire for thirty minutes or so, then hit the sack around nine o'clock or so. I had noticed while gathering some firewood earlier that I was still kind of weak in the legs but blew it off as part of the hike up the steep eight-mile hill yesterday. Around 11:00 p.m., I woke up to the sound of gurgling in my lungs. I was mildly concerned and thought I better stay awake and keep an eye on myself; around 3:00 a.m., I was now violently coughing up blood. My loud coughing woke up Eric, who was in his own solo tent thirty feet away; he asked what was going on and if I was okay. I told him flat out, "I have edema!" He asked me if that meant we needed to get out of there and get to a lower elevation. I said yes, but the first two miles of this trail going out is rough and in the dark would be very dangerous; it would be very easy to get a sprained or broken ankle or worse! I told him that I was still pretty clear in my head and that we should monitor my problem and try to wait it out till sunup; he said okay! Around 6:00 a.m., he got me up and out of the tent; he was stunned when he saw my face. He told me that I looked dead and my color was stone white. I was responsive, but I felt very far away from where I was. Eric said he would pack up camp and get us ready to leave while I ate the half a bagel he'd pushed into my hands. While he got busy, I choked down half of what he had given me and had a few ounces of water but couldn't take any more; I felt like I had been drugged! Eric examined me again and said, "That's it, I'm gonna set up a tent, put you in it in a sleeping bag, and run out for help." I flat out refused to be left; I told him that I would be dead by the time a rescue made it back up there. Desperate, I told him that I could make it out and that I would carry my own pack too. So we got me loaded; then he put on his pack, and we headed for the trail fifty yards away. I barely made it then crashed to the ground. Eric appeared very worried, saying I couldn't make it. I argued that it's downhill virtually all the way, that I'll just lean hard on my trekking poles and I'll make do. The reality was that I was in no condition to be making any decisions for myself, so Eric had to set a pace that I could handle. I would take about twenty steps; then he would sit me down (physically or verbally). My eyes would close, and then I would fall over, that is, if he didn't grab me first! At one point we even tried giving me some GU (quick energy food), but my condition didn't improve. It was amazing that on this particular trip, we had only seen one other group up here in this popular place, and they were long gone. It took eight to nine hours for Eric to mush me outta there; then he had to drive me to a hospital several hours away. When we got there, I was checked right in and put on oxygen while Eric called home to tell his mom that her husband was in trouble again. My oxygen saturation at 4,600 feet elevation was 72 percent; the doctors figured it was even lower while I was at the higher altitude. The staff couldn't believe that I had hiked out of Brown Duck and was still alive. I was asked if I remember anything, I said I didn't, none of them were surprised! I stayed in the hospital for several days and was visited quite a bit through all hours of the night by the medical staff! Around 2:00 a.m. the first night, a doctor came in to check on me and told me that he was a climber and asked a series of relevant questions concerning my activities, my acclimation claims, how stressed I might be, etc; after I answered his questions, he told me that his best friend had died after a trip to the Tetons the year before in a near identical situation and that there had been a big resurgence in altitude-related injuries that were more or less not making sense. I told him that I was surprised at how fast it came on and how quickly I had gotten out of control of myself, and that I was surprised I was still alive! He then added that if we had waited much longer, I wouldn't have made it and that I was still not out of the woods! My edema turned into pneumonia, and I was put on four liters of oxygen per minute for several weeks; in the end, my lungs suffered some permanent damage as well. What a great backpacking trip that was: I scared the hell out of my son and my wife and me and now had to return to my job to see what fate awaited me and my comrades.

My first week back to work, Joel and I got stuck cleaning out a storage building that contained at least two inches of dust standing on the floor. The masks we had been given were the paper ones that maybe help a little, but the dust penetrated these masks and our airways easily, making us cough and sneeze, and our eyes watered like crazy! Then we were assigned to repair a pump located in a 3,000-gallon storm well that had overflowed during a big rainstorm. As it turns out, the pump was a swamp cooler pump that could never hope to drain such a huge tank. Joel and I talked with the local city storm drain folks who agreed to come and pump our well for free because of the National Guard's service to the nation. When we finished the assignment, our so-called managers separated us for good. Lately, our foreman (the brute) had also finally begun preparing our performance evaluations for 2009 during September 2010 (he was only half a year late) and rated Joel and several others very low, actually substandard, which is the beginning of the end, especially if you can't appeal the rating, that is, to have HRO look into the facts and actually do something. Joel decided to cut his losses and retire two years early, leaving around the end of November of 2010. I was very sorry to see such a skilled guy leave and felt very alone, but it really was for his own good; his feelings were hurt with the low scores on his review, and he felt humiliated! Joel and I had quite a bit of fun together! We did astronomy after hours on some of our road trips, found the healthiest places to eat, our properties were in a great shape, and we had developed a good relationship with all our customers. Our goal was to take care of them, and they sent it back our way too, oftentimes inviting us to lunch or dinner, even an after-hours scorpion race that we chose not to enter.

When given my rating, mine was low too! I had received a zero in technical theory and hands-on capability; it was such an insult! I decided to go to our federal legal office for an opinion and was told that when you get a low performance rating like this, you will be given one more chance to correct the deficiency; then you could be terminated. Such an event could possibly interfere with any retirement benefits that I hoped to gain from this employer, depending on the infractions at the time of termination! So I went to HRO and met with the director and filed a complaint. I stated that the individual rating my performance and skills has an apprentice HVAC skill level and has no idea what my job entails, plus I have been a licensed electrician for over thirty years and have proven myself worthy of the license over this span of time. I demanded to see even one failure that was ever reported concerning my job knowledge and proficiency and performance. I insisted that I would be happy to take the state exam again, but warned the HRO that if I had to go to such extreme measures, that I would be forced to take legal action against our office. I asked the director to realize just how stupid they are going to look in court when the judge finds out that I was dismissed because I had no capabilities. Yet I had endangered people with my work for three decades and had received outstanding performance evaluations over all that time, including excellence awards, bonuses, etc. I wonder who will win on that day in court? I also challenged this rube with the fact that it's my right to be given a chance to succeed. Now the dummies have to train me, so why on earth did they hire a zero-qualified electrician anyway? Dumb, dumb, dumb, dumb dumb—this has to be a song, and maybe a dance! I reminded the director that this evaluation will go on my record if it is not corrected. He said, "Don't worry, you'll be fine, just keep working like you do." I felt like I just got the used-car sales pitch from hell. Finally it was New Year 2011. I had lived to see another happy freakin' new year! I was beginning to wonder what was so great about life. I felt like my remaining coworkers and I were in a Turkish prison of sorts! All I wanted to do was serve the troops and help them succeed until my time was really up or when I had to retire because of injuries. But how was I going to survive these killers, how was I going to survive me? Lately, looking in the mirror was not a good idea; I did not like the view, I was haggard and unhappy. I decided to get back into field testing and went to the winter trade show held at the end of January; I needed something positive in my life to anchor to, something to look forward to.

I found my buddy Jeff Lowe in a wheelchair succumbing to a version of ALS (amyotrophic lateral sclerosis). Here is a guy that just a short while ago was a leading American Alpinist who had always raised the bar, setting new standards all along the way. He and his brother Greg had created and sold all kinds of climbing and backpacking equipment for well over three decades, helping outdoor enthusiasts to more fully enjoy their sports; their company was known as Lowe Alpine. It was another day of reckoning for me. As I chatted with Jeff, he commented on how healthy I looked and pointed his finger to God, saying, "We know who saved you!" This was very humbling considering all the damage I had incurred from the lightning bolt back in 1990, not to mention all the subsequent surgeries, including rotator cuff tears in both shoulders, and now the edema and pneumonia. It was a lot to deal with, and all of this is notwithstanding my job and the deadly stress presently being dished out. But I am upright and very functional compared to Jeff, who was supposed to be gone by now; I was lucky, so very lucky! And yet here's Jeff in a wheelchair, directing traffic for his Metanoia project, *The Adaptable Man*, a film about this great climber's life and a solo climb of the Eiger, which changed his life for good! Nothing gets this guy down, he is truly amazing! Obviously, he had far more taken from him than me, and yet despite his disadvantages, he finds the way around the challenge to reach his goals. So you gotta keep busy and dream big and have faith; things always change! You gotta keep finding a reason to live and keep adding to the common good like Jeff; his example helped me to press forward! From Jeff, I learned that hardship is a normal state of affairs—it adds to your abilities to cope, you simply have to press on; if you give up for a second, you're finished.

Like Jeff, I spent quite a bit of time in the mountains around Utah and Wyoming and a bit in Colorado. I had lived and breathed rock and ice climbing; I also backpacked and learned quite a bit about the tools of the trade and the various brands of equipment available long before I ever field-tested professionally for the manufacturers. It was a total blessing that I met Wayne and was given the opportunity to test for him and the countless other companies he helped me become acquainted with. I met hundreds of people in the outdoor industry and used these opportunities to tell my lightning story and get acquainted with them. I gave my left boot (the one blown off my foot) to Asolo to display in their museum in Italy—complete with a synopsis of the situation. These days I carry my sock, pants, and my hat and have photos of the boot to show interested folks. While at this show, I noticed that it felt good to be there, and it felt good to know that people were still interested in my feedback; it was good to know that somebody valued me and my opinion.

I say this because of the rating my foreman had given me on my performance evaluation. I realized that this was a ploy designed to upset this proud electrician and cause me to leave the National Guard on my own. These actions and the complaints from other departments were known and were being followed by HRO and the legal office and ultimately our general. But my partner, Clark, who had been hired for his great skill with electricity and getting along with people (especially unruly military types), finally had had enough too and retired early like Joel and for the same reasons. So now I was alone again and being pounded by the stress and threats of the environment and the workload, which included my own desire to tough it out and serve my military brothers. At the end of 2011, our politically-oriented colonel retired. The general consensus was that he had spent too much time politicking as opposed to being fair and doing his whole job. In some ways, I don't blame him; he has a huge job to do, and his deputies are supposed to handle many of these issues, which he had inherited when he came on board, but ultimately it's his problem. Our general was also looking for change, realizing (I imagine) just how dysfunctional our department really was despite or in spite of all the management changes. So our coordinator moved along too, filling another troubled coordinator's position at the Air Guard with his own inept style, despite the fact that there were at least two very qualified applicants who had applied for the position; he lasted just over two years. Can you see the vicious cycle we create when we move problems without addressing them? So why do we have management processes, when we don't use them.

One day the quasi criminal may outnumber the average citizen, and the balance or scheme of things already may be in their favor; after all, they are very powerful, and they appear to have our government on their side, and it ain't going away anytime soon! You may or may not agree, but I consider most of these otherwise normal folks to be criminals. Sure they're not in gangs per se or part of terrorist cells or by definition are they a semi-new version of the threat; they seem to want similar things. No, I don't think for a second that the perfect world is ever going to happen, but we all could be a little more honest and follow the rules that are supposed to apply to everyone! And no, I don't think we should be dead rigid either, especially all the time; there is a time and a place for everything. And just so you know, I'll tell ya, if there is a judgment day, it is going to be a big mess—worse than hell. The godhead better ramp up job service right now so they'll have a big enough staff ready to handle all the issues! Jesus and his apostles will be on overtime as it is, runnin' around like chickens with their heads cut off. And us? Well, we better have fully charged "Kindles" ready to help us while away the hours. 'Cause if the military taught me anything applicable, it was about waiting, and wait we will for our turn in the barrel to see who tunes harps and who feeds coal into some boiler down below.

Well, the watchers caught on to the phonies and made some changes even though it took a while. I was skeptical and concerned as it appeared that I worked in the "mother of all black holes"! Our general replaced our previous leader with a colonel who was known for doing things the fair and right way; he was known for getting all the facts then acting on them. His marching orders were to finally get control of the maintenance office and bring it up to required standards, which would necessitate some intense focus and deliberate calculated actions. Did he do it? Kind of… It seems that all the previous colonels had been placed in this position as a short-term assignment until they retired from the service, usually within several years. None of them seemed to have much interest in correcting status quo unless forced to. I'll bet they all wondered as a whole what they had gotten into. No matter what, I think a person should not worry too much about his last day in any activity till several days after he is finished. Putting your best foot forward—whether it's your first or last day on earth—may well be what you are remembered for.

Per our new colonel's request (he knew my history and trusted my judgment), I gave him a ten-point bullet list of what to fix first over the course of one year's time; then we would work on new milestones. I asked him to be sure to bring in other employees—past and present, including several vendors—for their input too. It was important to keep the process well-rounded and productive if we were to have a fighting chance at repairing the mess, which, in my opinion, was somewhat less productive than a post office at Christmastime, where the waiting line can seem like death row. I knew this process would give this former MI (military intelligence or "mentally incompetent," I never figured out which) officer an eye-opening accurate view of the facts so he could proceed in a logical and fair fashion. The colonel during this first meeting asked me if I trusted him to do the right thing; I simply and flatly said no! At my response, his face flushed white, he looked shocked, then after semi recovering, asked me why—adding, "You know me, right?" He told me that even though a big chunk of his experience was in commo or telephone services, he understood the maintenance requirements and the required pace broadly. So I simply again said, "I have been here—it would seem—forever, and I have been involved with the government and the military longer than forever."

I keep thinking that I have been to the mountaintop in bureaucracy and self-serving processes; then a whole new level begins, and the bottom falls away! "Sir," I told him, "I have become very careful of what I believe will happen now or in the future here, except I know that it will be different in some way and will most likely rest again at status quo." My new colonel was aghast at some of the input I gave him concerning the goings-on right under our general's nose; just the customer complaints (the troops) ought to raise eyebrows and questions, why hadn't they? Missions had been at one level of risk or another because of some of this baloney that we as a whole allowed. These discrepancies had been hidden behind words like "everything is fine" or "we'll get to it"! And since there were no real checks and balances or accountability, the mud flowed. I told my colonel that I wanted to have faith and would exercise just a little, but in the end, I told him, "Prove me wrong! With your permission, sir, I will check on the current milestones you will be going after to see how you are doing and will make comment as we go"—just like a good master sergeant should do! He welcomed the assistance but could not fully understand how used up I really was; I am not sure if I understood either.

The blinders I was wearing even hid my failing body. I was very out of balance by this time and heading downhill toward a stroke; it was hard for me to believe that! Ever try telling a drunk that he is so, he will fight ya tooth and nail and complain that you are the problem, not he. Our new colonel began by surrounding himself with a proper support group (i.e., Chief Duke and Colonel Luke). Then he stepped up communication to the group and took affirmative action to gain some immediate organization; this smelled pretty good and looked good too, so I thought we will see just how it tastes. Our colonel routinely told me that he was busy with our bucket list. He knew that with my rank and dedication level, I could be dead and I would still be watching and complaining and directing. At my suggestion, the colonel moved the rude foreman to an isolated operation on the west side of the valley, getting him effectively out of management, and gave him a little different set of responsibilities to attend to. I think the colonel was surprised I didn't recommend immediate termination. But as I explained to my leader, the Guard allowed this individual, as unqualified as he was, to be hired, then kept him on the payroll long after he should have been removed or moved. As I see it, it was only fair to give the guy a chance to succeed at his former trade, but he should never again acquire the position to manage people since he is by nature better at abusing them than leading them. My gripe with the move was that the new leadership did not tell the foreman the real reason he was going to the new assignment, and I saw that as a mistake. You need to be upfront with people—even mean ones. Eventually, the colonel did have to level with the brute and deal with his nasty demeanor.

I found the changes to be coming along, and they were good—a bit painful and slow to materialize, as is often the case when progress is happening. I felt sorry for this colonel for having such a harsh detail thrust upon him, especially after just having returned from a tour of duty in Iraq, and I told him so. Cleaning up long-term messes like this one can eat away your soul; and you may not, for many good and bad reasons, even accomplish the goal. All our colonels had been faced with this issue! Insidious negative attitudes can completely kill the effort, no matter what's at stake. It's as if, like I said before, that we have built in terrorist cells that eat lunch with us, type letters with us, and spend the day with us only to worm their way into the system and bring it to a halt. Maybe they don't mean to be like they are by and large, but individually we really have to ask ourselves the question, just what the hell are we doing, and where do we belong? It's the rare individual who stands in the gap and sees it through, that is, as long as he or she is allowed to.

A short while later (early 2012), our new colonel hired a new coordinator, Peter; he and the detractors that were left over from the previous coordinator went at it right away! I am sure he felt cornered by the hungry rats. Immediately, he let me and a few others know how valuable we were to him; in countless ways we reciprocated that respect as we formed a bond that actually felt as if it had been there for millennia. One day, out of earshot of others, Peter told me that now he had to figure out a way to keep me long-term! Apologetically I told him that he would have me for about a year! I think he sensed how ate up I was. The damage brought on by the previous management ran very deep within me just as it did with some of my other coworkers; we were prideful people. Even though I knew good things were coming, in my own way, I was dying; and the vortex of retirement was pulling hard at my being. To me it's so amazing that when it is your time to change, it happens, and nothing can seem to stop it; inside, down in your soul, you know it is right, and it must happen—the steps must proceed to that day no matter how much you may not want them to. Life is so short, and I have some other things that I must do before it is too late; one of them was to write this book. I also have to take care of my family and be more associated with them than I had been!

Internally I knew that my time was up; my neck issues with all of its dozens of fractures and nerve issues were again becoming a significant challenge. My doctor told me that with all the hyperextension that my job demands, that I really ought to think about calling it quits. I am sure that all this weighed heavily on my wife too, because on both sides of this fence, many questions about our future were raised, questions you hope you can attend to and still survive the outcome. I was afraid again, feeling cornered, I really needed to stay in life and again had not really thought about the day when I had to quit my new version of normal life. I guess I had done a pretty good job of pushing all the issues that the accident brought off to the side. But now they were, at least the big ones, on center stage—the same stage I had fought long and hard for, cried over, tried over, lived for, and gave up for. Now the game was coming to its logical end. I guess I had fought a good fight, and ultimately the decision rests on me, so I would do the smart thing and listen to the whisper from my soul's caretaker.

In the early part of 2012, I began to receive injections in my neck column to help reduce the effect of the deterioration of the L1–L6 vertebrae. The spinal institute I had gone to before was now having me see a PA (physician assistant) instead of my surgeon. I didn't like this idea as the PA didn't seem to listen, and I had absolutely no history with him. Apparently I had to see the PA first; then I could have a conference with the surgeon (I found out later that this was not true). On the day of surgery (when I went in to get my first injection), I was given a consent form that talked about the very real possibility of this process killing me and/or paralyzing me, along with a small list of less significant side effects. I was a little miffed that this was not discussed before this day! As I entered the operating room, the PA and the surgeon who would actually be administering the injection were locked in a debate about at which level the injection would take place. The PA was opting for the highest insertion point, whereas the administrator wanted to insert at a lower place in the neck making it much safer and less likely to actually do me harm! The PA insisted that the higher they insert the device, the more likely it is to work, and if we go lower, the injection may not work at all. I butted right in complaining that I should have been alerted to the dangers of this process long before today. If an injection was going to take place, I wanted to approach it from a reasonably safe standpoint instead of a high-risk one. If it doesn't work, then we move on and talk about the full-plate installation that would replace the existing half plate.

So I was injected at a lower site, and even despite my high-activity level as an electrician, this injection brought me comfort for the better part of the year. However, my job with all the bending and articulating really beat the crap out of my spine—my doctor (the PA) again told me that I was really past due for the full plate and should consider doing it soon. I could see that I was about to become even less mobile than with the first plate and wanted no part of it. Besides, the first plate's surgery had left me with some serious anxiety that for a long time made me feel as if I was choking to death and amped up my stress. I know this was a state of mind, but it was real enough for me. Besides, if there is a less dramatic approach, I say take that road for comfort's sake, manage yourself more carefully, and maybe gain some quality of life in return and still be useful to somebody. If I got the full plate, I would have to retire—like it or not—from the electrical trade right away, and I knew it. But I wasn't ready to be forced out if I could help it. I wanted some say-so in the when part. I wanted to say that I left the final battle with honor having served my time in an honorable and respectable way while serving others, including the mission. I did not want to leave because some knot head ran into the room and ratted me out saying this man's sick or incapable, here's the papers, see! Hell no! It's bad enough that many of us were trashed and kicked out like dogs, like my partners and many others over the decades, never to be heard from again, their individual and collective service ignored; but again, change did come!

Changes to our overall operating methods began, things that I and the other two amigos had initiated long ago—man, it was painful and stressful, a bit difficult even, and was actually quite similar to starting up a brand-new business. The other two amigos, because of command directives, added office staff to help instigate a quick turnaround in the way we operated. But sometimes, there is so much wrong that you have to approach the elephant with small but focused priorities and start eating a leg, any leg, and continue till you've finished. That done, you get to chompin' on the next most important leg and so on—it's how you eat the proverbial elephant (sorry for being so cute) one bite at a time. But the previous management had really done a job on our operation, costing the taxpayer hundreds of thousands of lost dollars through the various avenues they impacted to the negative. The previous coordinator, Kent, had even thought it a good idea for OSHA to come and look at our facilities (without our commander's permission); he wanted a friendly visit. I told him that I had been an electrical inspector in my past and that this facility "headquarters" was not the place to let OSHA have at it because it will not be friendly; it will be costly and that we should find a more productive means to overcome our safety issues. As luck would have it, I got to be the escort for the OSHA person who looked like she was born with a chip on her shoulder right from the start. I requested the friendly visit as advised, which seemed likely until our inspector noticed the first of many blocked and nonfunctional fire doors, stating, "Well, that's the end of the friendly visit!" This one infraction cost us thousands of dollars in penalties, now we had just several hundred more infractions to answer for. I wondered if our new coordinator was just plain stupid; he could have found help through Risk Management and avoided the impending OSHA showdown, which he invited! He did not follow protocol and should have been reprimanded. Military security had also become a beast of sorts too and is very strict, so much so that blinking an eye wrong may bring the secret service down on you just to find out what you meant. Background checks became very extensive. Searing, prying, redundant appraisals of your right to be an employee with the government can take months and thousands of dollars to complete; you can thank the Taliban for this affliction. These new requirements really made hiring people very difficult. We considered calling the pope to see if any of his comrades would come to work for us; surely they might have a cleaner background check—you know, no parking tickets and the like. Buying parts also became very involved from a paper trail point of view (say good-bye to the trees), which also included an electronic trail too! Like it or not, some resource has to be used up, but the inventers, they don't see it that way. They think electronics is easier, cleaner, and cheaper; the truth is we are currently, at HQ, losing a minimum of one to two hours per day of productive time because of computer work–type red tape. But that's a management issue that could be fixed by using secretaries to deal with some of these processes instead of maintenance folks. Maybe they'll get it figured out, or it will right itself and everything will become smoother? See, I'm fairly confident in the system. However, the new maintenance module computer program has been wasting our time and energy for half a decade already. What I really think and know is this: some managers, plain and simple, don't know how, or care enough, to organize a maintenance crew. Sadly, it is so easy, but some folks prefer the "square wheel and excuses" approach anyway.

I swear the government purchases delinquent programs as a routine wasting millions or billions while trying to implement them, then winds up with something newer as a fix; maybe it's the reprogrammed version of what they already had? Nevertheless, the vicious cycle of square wheels starts all over again. I wonder, did it ever quit? My new leaders (the colonel and Peter) needed me to perform like the self-proclaimed world-class athlete I used to be! But I had become—not burned out by the trial by fire, but rather burned *up*, and felt as if I were gutted and dying! I had used up my reserves, and now I had to leave—really, for all our own good.

This next year, my last, was incredibly busy. Peter was trying his hardest to keep up with all the demands of the customers to include repairing what the dynamic duo (Kent and the brute) before him had screwed up! Man, this guy had an amazing amount of forward-moving energy! He also battled the detractor crew members and HRO as a routine part of his workday. He dutifully responded to the high op tempo demands of the new colonel, who wanted to see big changes right now! Unfortunately some of this pressure came from me too, as I was routinely checking up on this commander to see if he was following the process I had given him to accomplish. I have to tell you that I really am surprised that he listened to me (I have never trusted an officer either). However, given his interactions with the previous coordinator and the brute, I am sure that my advice mixed in well with all the other folks he had the occasion to interview, which caused him to realize how correct the angle of attack was. As a matter of course, he would see me at times and tell me things were moving forward; I played the hard guy, saying, "We will see!" Peter on the other hand even went out and did maintenance work himself or teamed up with one of us maintenance folks to get through one problem or another—now that's correct management. This guy was in a serious pressure cooker that even led to a dispute between the detractor workers and his wife, who worked in the communications department. Let me tell you what, though: people with self-serving agendas will always find a way to have their way; just like criminals, they always return to the scene, and they always repeat themselves, so they are easy to derail—if the system governing them is not broken. Like I said before, we have made rules to live by in society, but there are those who don't want to play by those rules; they have their own agendas, and this hurts everyone. In life, we all need to stick together so we all can have meaningful happy existences at work, at play, and at home; and maybe we can even have a little balance in life as we grow up or old. A buddy of mine once said that all he expects is mutual respect and common courtesy—there it is, nice and simple!

I don't think our military leaders recognized what was unfolding right in front of them. The detractor workers were getting their way by making Peter look like a bad manager; it must have been terrible for him! All he wanted was to see the National Guard show up on the map and be accountable for itself. He left after one and three quarters' years of service. I will say that in the year that I worked with him, we accomplished a hell of a lot of work together, saving the taxpayers lots of coin and can be proud of all we did for this organization. We worked real hard for our folks; it may appear that they don't notice, but they do, and they remember the good and the bad, so we tried like hell to do no bad. I am sure that in the future, Peter will be missed.

Anyway, as we went into the fall season of 2012 of my last year, the workload really increased! This time of year is known as year-end spending, where all the extra money left over from that year's budget (one of several) gets spent on whatever.

Another electrician had been hired by the previous management to replace Clark, but didn't seem to do very much work as he was always fighting with Peter about what his job was. He claimed that he was hired to manage projects and was the new lead electrician (my job). He was being paid eight dollars an hour more than me to fight with the new coordinator, Peter. Instead, he should have been learning from me about how to function in this runaway locomotive environment. To be fair, he is a very capable electrician and had come from a new construction environment, the shift to a maintenance/military environment takes some getting used to as you are dealing with repairing, replacing, and maintaining things as opposed to simply installing something that you didn't or couldn't design. Also, I worked in multiple hi-tech areas of the electrical career field designing and installing special circuits for special equipment, something your common sparky never gets to do.

As 2012 leaned toward history, I really got hammered physically. I really don't think I have ever worked so hard in my entire life, and at such a tempo that time became meaningless, but the pain was for real! One evening during September, I think, Chief Duke told me that he and a mutual friend (Glen) thought I ought to check into the URS (Utah Retirement System) and see where I sit. Actually Duke thought I had done enough work anyway and should go enjoy life. So my wife and I scheduled a meeting with the retirement office and found out that I had over thirty good years with the state of Utah collectively with my service to the school district and the National Guard combined. We learned that that we could retire with only a small change in pay! My military retirement would be kicking in three years from now, which is when I turn sixty, and would help make up the difference. So it got me thinkin' and it got my wife to thinkin'—remember the spouse at home suffers too when the significant other doesn't come home till late evening. My wife had been very concerned for me for a very long time now and saw a way to get me home and in just a few pieces as opposed to many.

Anyway, the other electrician, like I said, had been hired to be the lead electrician by the former coordinator, Kent—well, I was the lead electrician; when I counseled with our HRO, I was told not to worry, that everything will be okay. Righty-o, pally!

So now we had two lead electricians on our budget; he leads me, I lead him—how silly can things get? It didn't feel quite like yin and yang and wasn't very symmetrical either. This other lead guy told Peter (the new coordinator) one day that the $8,000 thermal-image guns that somebody claimed we needed to do our jobs (*not*) would take weeks to program. Peter needed to get them ready so we could teach the staff over at Camp Williams how to use these wastes of big money! Eventually the boss came to me asking if I could handle the programming; it took all of twenty minutes, and I'll bet my ten-year-old grandson, Bridger, could've done it in fifteen. Peter wasn't shocked or surprised; this other electrician had not come here to work, but to manage contractors!

The detractors were determined to completely restore the facilities with new equipment and wiring inside and out, top to bottom, and purchase high-end useless tools to boot—all on their own authority. Nice idea, I admit it, but we are a maintenance operation; we preserve our resources saving valuable dollars by maintaining what we have in good times and especially in hard times, like when money becomes scarce and jobs can and do go away because of funding problems, and yes, we should take care of necessities! Obviously there were gross misunderstandings all around.

Well, time was now going away so fast that it was bending light! Myself and one of the other HVAC guys, Brian, helped each other quite bit and struggled together as the weather (literally) worsened. We presently didn't have proper trucks to go do our jobs (remember, they had been traded away by the smartskies). So as it rained and snowed, our tools and parts would get soaked; we tried using plastic tarps to help, but it was miserable anyway. The maintenance module program was eating up our time and aggravated anybody who really cared about getting their work done, which were really only a few folks. Some of the detractor workers (a.k.a. Snipers) were more interested in driving around, trying to get the rest of us to slow down and/or not complete work, ultimately to go against the Guard and the new boss—it was amazing! I told Chief Duke one day late in that year that our department was about to repeat itself, that the new crew (detractors) was more skilled than the first crew we had worked with years ago, but they were of the same mind-set and that they were winning the internal war with the Guard. He agreed that their MO (method of operation) was shockingly similar to the ones of yesteryear and that the management was poised to fall prey to them. This happens because the management is very busy but also very disorganized and simply won't spend the effort needed to bring about order—so it's "status quo" in the land of the free all over again. In my heart, I felt a coffin nail sink into the wood my head is made of.

November and December were all tied up with modular furniture jobs located in large managerial departments. These rooms are well over a hundred feet along each wall with fourteen-foot ceilings containing up to a hundred individual light fixtures.

When we change to the latest greatest furniture due to excess funds at the end of any given year, we displace the local workers for several weeks at a time while the remodel happens.

It never fails: the IT folks usually have a new design of cable to run, and I have to reestablish connection points for dozens of electrical circuits, and I make sure that any new or larger load are accounted for, and it all has to fall together in some semblance of order within about a week and a half. This particular year, I realized that the furniture people had really cut back on their engineering staff. They were employing people to design their systems that really didn't have any electrical background; this means that circuits were being overloaded. These designers were placing the connection points in poor locations and didn't ask how many circuits were available! Also, the quality control of an installation that goes up like a wildfire is fairly bad too. With modular furniture, it is imperative that the internal electrical junctions are solidly connected. You see, the plug-in points for these connections always fail top to bottom, meaning that the common or neutral wire disconnects first. If this happens and the hot wires don't disconnect, you wind up with a voltage that is double of what it should be (i.e., instead of 120 volts, you wind up with 208 volts or thereabouts, which will burn up any 120-volt equipment plugged into an outlet requiring this lower voltage). In the best-case scenario, all conductors are disconnected and no harm is done and is nothing more than inconvenient! These projects have become so fast-paced that you would swear that there was a Black Friday sale at Walmart and some smartsky yelled, "Hey, free TVs!"

One day my electrical buddies from the Air Guard, Mark and Eric, had come to HQ to a health fair and saw me in my usual big hurry; they both said I looked way too busy. I took Eric into the USPFO office space that I was working in and showed him the immense project that I was doing alone; he was shocked and couldn't believe that my portion hadn't been contracted out. When I told them about the short time line of the installation and associated troubleshooting, they were immediately happy to be working where they did. Actually, a pair of skilled electricians can do amazing things, especially if there is reasonable planning and time allowed to do it in. It's not fair to stack job upon job on some guy day after day with little room to think or plan—being a machine gun can be fun, but how long can one remain effective without proper maintenance? My second-to-last job convinced me that another nail had been driven into the coffin—troop command; another large office space, and for no good reason, had been planned by the military to happen just before and right through the Christmas holiday. Perfect, I don't even get a break even though I had applied for leave during this time months in advance. Typically nobody works during this holiday unless it's an emergency! Brian, my HVAC buddy, helped me quite a bit on this project, bless his soul! This project had its share of fubars (fouled-up beyond all repair) by design, which I had to rectify with the electrical designers who didn't even understand common electrical terms like *volts*, *amps*, *watts*, *volt amps*, etc., but I helped them help me to be productive as possible so that the finished product would be correct. I sensed that they were intimidated a little by my presence, but with electricity, you gotta get it right. To complete this project on time, I wound up working right through the department Christmas party; many soldiers and civilians wondered why, even our general did. It's okay, I filled a need.

Being a stickler for detail, Brian and I had corrected quite a few connection issues that the furniture installers fubared. We also had labeled circuits, power poles, and had created one-line diagrams of the electrical part of this project and felt very happy with the result! You see, Headquarters has a lot of electricity available at the three services we own! One is 1,600 amps, and the other two are 2,600 amps—respectfully and collectively, they feed over two hundred panels, which mostly contain forty-two individual breaker spaces unless they are distribution panels. We also have motor-control centers feeding boilers and big chillers too. Many of the smaller twenty-amp circuits have a purpose but typically use very little current; quite a few have even been abandoned. I found the building easy to manage if you are a thinker and have some skills, but this building changes like the wind in terms of remodels. Its needs on one of the 2,600-amp services really never changes; the other one has always ran at about 700 amps, and the 1,600-amp service is isolated for a special group and maintains a relatively average level of use—probably for hidden hot tubs! I guess what I am saying is, if you plan your work and work your plan, you can succeed; but to the untrained eye, it seems an epidemic is just waiting to break out all over the place—nothing could be further from the truth.

Through this time frame—December, that is—my wife and I visited the Utah Retirement office again, and things just seemed to progress as if by nature; there was no pressure or hesitation to say yes to that plan or this amount. I found the contrast amazing because my personal workload was planned out well into the next year, 2013, and I was ready! I had parts acquired, plans drawn up, hamburgers ordered with fries and chocolate shakes. During this visit to the URS office, Teri and I had to pick a day that would be my last day and then sign some forms and go visit a few other offices, one of which was to get my medical insurance in order. Every office was so accommodating; it's as if they had been waiting just for Teri and I to show up so they could help us evolve into the next phase of our life. No, I did not feel any vindication! My spirit seemed calmer, its whisper more clear than usual. Later after I had retired, I did feel like I had escaped an impenetrable prison of sorts, a place that had caused me to change a lot—I still wrestle on occasion with the idea of—was it a good change or a bad, change or a mix? All I know is how right retiring felt—just like when I first met my wife, Teri!

As much as this job helped me, it hurt me too—in some ways, just because of association! Throughout the state system, this place is known and disrespected for its work ethics, including its general ethics too! This moniker is really too bad because there are some real standout citizens who work and seemingly live there, which I believe represents the most of them. Like me, they are judged by the acts of the few, not the many. Most of these people got to see all my moods (us redheads are a pesky fiery bunch), and I saw theirs too. There were days when we ragged at each other, but mostly we supported and complemented each other as a routine. I've come to really like these folks I served, who are the same ones who serve us with their repeated tours of duty all over the globe. I am proud to say that I value highly my association with these army folks and show it through salutes and the wearing of the pride hats and shirts they gave me bearing the Army National Guard logo.

The plan was I would give my notice on January 2, 2013; then I would work up to the fifteenth of January 2013, and then I would be officially retired on the sixteenth. I could rescind my decision (that is, change my mind), but I would have to produce a letter of explanation—not later than 5:00 p.m. on the fifteenth of January 2013 to the URS office. When I got to work on the second of January 2013, Peter and Sgt. Profit (one of Chief Duke's subordinates) ran the lineup meeting as usual. When it was my turn to give a short detail of my upcoming workweek, which would involve another big project in our SCIF (sensitive compartmentalized information facility) area, I added that I needed to talk with them in private afterward. After the crew Left, I closed the door then sat down directly across the table from my boss and Sgt. Profit and just spit it out, "I'm giving you my two weeks' notice, I am retiring, and my last workday will be the fifteenth of January 2013, and the official day will be the following day, the sixteenth." They didn't show it, but both men were shocked. I told them that in my mind, I wasn't retiring, I was quitting! I then spent a moment or two lecturing them about "duty to the nation" and staying on track and how much I appreciated everything they had done for me. As a retired master sergeant, I encouraged them to pass the advice on ethics and military bearing up and down the chain of command. I encouraged them both that this decision had nothing to do with them and that I was sorry to be leaving the battle at this point and that I would work hard for them right up to the final minute. I went on to say that up until the recent past, this job had been the worst experience of my entire life! I told them about the possibility of rescinding the decision and its requirement, which would become a mute point at 5:00 p.m. on the fifteenth of this month. I also apologized to Peter for not talking with him sooner, that I didn't want to ruin his holiday. I let both men know that I had considered changing my mind ever since I had signed the papers at the URS office in December, telling them that my wife and I had engaged in many deep conversations about that idea. I had done my best to make a decision from a calm, realistic standpoint as opposed to one driven by my anger concerning various historical situations that would never be reliably fixed, so to speak. I mentioned about how skeptical I was in this light and my trust meter was pegged at zero—with the needle bending. I told them how my wife and I become involved in the decision (Chief Duke's suggestion) to even check out the process to see what the possibilities were, and that we had been deliberating for several months since. I also told them that my doctors had recommended that I at least change professions for the good of my health. I had recently been informed that my neck, at the present pace, would soon dictate a change anyway, and I may succumb to a permanent problem like paralysis. Focusing on Peter, I related that all the work of late had just been killing my neck, and I knew it was past time for some more help via another injection. I had started going numb again down the left side of my face, neck, and shoulder, and the pain was getting unmanageable at times, and I wished that the other electrician could have been more useful. A lot was on the line on both sides of the fence! I really didn't want to leave until I was sixty years old because of the seamless way I would gravitate to full retirement between the state and the military, which had been my and my wife's plan for some time—but again, things had changed! After I was done, Peter and Sgt. Profit said okay, and we headed off to our duties for the day. I entered the SCIF and began my last job for the National Guard.

In the SCIF area, you must have an escort, and I lucked out by drawing the commander of that operation, Major Schnell; he would babysit me all through the project. We were friends and talked at length while we worked; yes, he helped! The major couldn't believe I was leaving and didn't know what he was going to do without me! Major Schnell complimented me on how resourceful and helpful I had been to his operation. I thanked him for the opportunity to serve him and his staff—such respect! However, as the job proceeded, I was winding up in very difficult and demanding positions with my body, involving mostly my head and neck; it was really killing me and was so painful that I would groan and react to sharp pains in my neck and head. This situation caused the major to offer to do the work if I would just show him how. I thanked my friend but told him that for the good of his mission, I must complete the electrical work! Major Schnell took me to lunch every day during this project, and we bonded a bit more as a result—food and work seem to do that!

During lunch one day, the major told me that I really had to stop thinking about the mission and my work here at the National Guard and start thinking about *quality of life*, which will involve not only me, but my exceptional wife, my kids, and the rest of my family and friends. It's a real blessing to have even just one *true* friend while you are in life, and I have many!

The SCIF job was extensive enough that it used up the remainder of my working time left with the Guard! On my last day with my friend Major Schnell, he told me that he too was leaving; he was relocating back east for a while to gain more skill and thanked me for all the work I had done for his operation over the years. I fired back saying, "But you guys had to put up with me and all my smart comments about Colonel Flagg from the Mash episode being your cousin." He laughed and said it was a real pleasure; we hugged and said good-bye. While I had been finishing things up in the SCIF, Peter was communicating with our command about losing their most productive moneymaker in just a few short days! The plan(s) they had devised to cause me to stick around was kind of humbling. By the time D-day or January 15, 2013, showed up, I had been offered the opportunity to rewrite my job description, and it was pretty open-ended—cool, I thought, but it also made me suspicious because the government has an office that does this very thing—and for legal reasons, obviously. I was a bit humbled, yet a bit amused with Chief Duke's idea that I could be in charge of an extensive preventative maintenance program statewide and would be my own boss and the primary worker. I would have my own truck or trucks and a helper, maybe! I would—again, just like before—develop a plan and a paper trail and design the complete hands-on aspect too; it did sound appealing, but I had to wonder what the other electrician would be left to do. This change included a month off—beginning right now, and more if I wanted, plus a minimum of an eight-dollar-an-hour raise to make up for the fact that the new electrician had be given a salary this much bigger than mine when he became employed by us, hmm. Interestingly this amount of money placed us well beyond the top of our pay scale that is achievable in a full career. From the standpoint of how the state retirement system works, this definitely had its advantages; in my case, a retiree's income is based on his three highest paid years of service, and I knew I could survive this place standing on my head (neck injury and all) easily for three years, especially considering what it could mean to my soon-to-be-realized fixed income! I guess I'm just too honest! I just know this, that if you pull out your personal history book then stand in front of a mirror reading out loud some of its contents while looking straight into your own eyes, that it may well become a very humbling and dramatic experience—one that I am sure would bring me to my own knees in the deepest of sorrow! I didn't need to add any more infractions to my personal list of issues, for my sake and God's! I also wondered why I had not been treated a little better before—no matter, because in reality, President Obama had cut back plans that could scrap this plan and others for our operation, which turned out to be the case. Plus, I didn't think they could really make this pie in the sky fly on a plate, platter, or saucer complete with all the little pilots, but I did appreciate the thought and offer. Other possibilities were mentioned that sounded pretty lucrative too, but it always appeared that I would be doing the lion's share of breadwinning! Now don't get me wrong, I don't mind hard work at all—it's exactly what I was built for, but I was tired of the taste of what I thought was unfairness and deceit for gain! I had become a little ornery as a result of this whole gambit that was being played out year after agonizing year; in the end, it was always the same old story! I'm glad that I was taught to be wary of things that were too good to be true, and I have realized all throughout my life that I should have learned to pay attention a lot sooner about the greener grasses that hide the bare, naked truth.

At 5:15 p.m. on the fifteenth of January 2013, the commander pulled me and my boss Peter into his office and said he heard a rumor that I was quitting and that today was my last day. I grinned a little, saying, that to be more accurate, fifteen minutes ago, it became official, and since I hadn't called my wife to tell her to turn over the letter of rescind to the URS office by now, that I was officially a freer individual; he seemed stunned—I know he is a busy guy, but he is the commander. So I told him that there is still one other possibility to change the course of the future, and that would be to have our general call the governor and ask him to give a directive to the URS folks to force the change. The colonel asked me if I was willing to stay even for at least a short tour if he did this. I asked just how valuable I really was to his operation; he replied, "It goes without saying!" So I told him I would stay but that I could not make this decision alone, that out of respect to my wife, she had to have an equal say in what happens! Of course he understood this because, like me, he had left his family so many times to go serve around the country and in the theaters of operation. Peter seemed hopeful and happier; I told both of them that I would have an answer in the morning. Needless to say, my wife and I deliberated again and for a long while, but in the end, we decided to move on. Now, it would be someone else's turn to grind out the answers and be diligent; my electrical days were over!

The next morning I gave the leaders the news and was congratulated for a job well done and spent the day tidying up my SCIF job, turning in tools and critical "as built" drawings out of my personal file. I wrote Peter a game plan or, in other words, my MO for keeping important things in focus and functional, basically an Air Force self-inspection checklist. I turned over all my schedules for the maintenance and fueling of our backup generators, which was probably the most important part of my job. I offered to come back and perform safety and equipment training for the department at my expense; I surely didn't want to leave them with any important questions open-ended.

I said "so long" to many a comrade, thanking them for the opportunity to serve them and thanked them all for their diligence and service. I know just how tough it is for the National Guard out in the world of the military. The Guard has a lot to prove and have many disadvantages that make their existence sometimes questionable and very difficult, plus they are fighting their very own history as a collective group.

In my own military unit, Electrical is just about the most technical career field there is. Asking a truck driver or banker to become technically qualified to serve in this capacity, receiving only a few hours of meaningful training a month, really slows down career progression and mission readiness; it's frustrating for everyone! But until the mucks in Washington eradicate these issues, we will do our best and press forward to become ready for action as best we can. If anything, the National Guard these days has a highly qualified full-time staff and quite a few traditional part-time staff who practice their career field as civilians. Us Guardsmen as a whole have big hearts and have evolved quite bit through the decades, especially since Vietnam, and have become a better help to the active-duty forces, or at least I hope we have, which is as it should be. We should never replace the active-duty folks as this is a huge mistake especially in an effort to save money (my nonhumble opinion). I have the greatest respect for my military comrades, especially the guys and gals that I served directly with; they are some of the best friends in life I've ever had, and I would go anywhere in the world with them—and I have. We took care of one another and were a team! We did ourselves proud with every job we did. I love you, guys! The civilian side of military operations however needs lots of focused revamping, period. These jobs are considered by many to be retirement ranches, but they are not; they are important elements of the consummate staff and are needed to help fulfill the mission. Hence my agitation with these topics!

Anyway, for me, my time this day was loaded with strong emotion, and as I made my way to the end of the day and my career, I realized this: I have stood watch a long time now as a traditional National Guard member and one of their few full-time civilian electricians. So despite my sense of duty, I now needed something more from life. The vortex of change had come and drew me away.

The Back Door

The next day, January 17, 2013, was a new day, a very different day, a rite of passage! When I woke up around 9:00 a.m., after sleeping for about twelve hours (a first), I felt great! I didn't feel any pressure—no stress either, not one ounce, no kidding; all of a sudden, I didn't feel like jumping off the nearest dumpster anymore.

I was told several days before by one of our officers that I was going to get up and drive right in to work today, like it or not—if not out of duty, then out of habit! The whole day went by without a thought of the headquarters. I didn't even miss my tools, which had been great companions along the way. My tools were great friends, and they always talked nice to me and me to them, even though I had (accidently) shocked a few of them once or twice.

Finally it hit me. I was free!

Then I thought, *Now what?*

Reconstituting back into life is very interesting, especially when there is nothing you really have to do any one day of the week except maybe take out the trash. This can get old and is something I recognized early on as a dangerous trend—I knew that I would need to stay active, and I was a bit nervous regarding the big picture of life that lay ahead of me now. I was concerned about relaxing too much—and what if I lose my purpose? They say people often die shortly after they retire. I wonder if people see themselves as not having a meaningful useful existence and, in every respect, fade away! I remember Major Schnell from the SCIF telling me that he hoped that he wouldn't be reading about me in the obituary a year or two after I left the army. I gotta keep busy, gotta manage the fallout from my lightning strike, and gotta stay positive, that's all. People say that when they retire, they will go fishing every day—maybe one in every ten thousand will, but everyone else, they will find themselves looking for something to do besides fishing, or they will stare at walls. I have found that as I progressed through my first year away from the mainstream, that it is important to occupy your time with some meaningful activities. At first, I thought about getting a part-time job at Lowes or Home Depot as the electrical tech. However, it must have been time for me to transition from being an electrician to something else because I simply had lost my desire to do that kind of work at any level, I had already done it forever, so I gave up that idea. Though I do think I may get a job, it would have to appeal in such a way so as not to impose too many limitations on my freedom and be something that is fun and interesting. Otherwise, I am happy with what I have found to make up my so-called workweek, which goes like this: I perform a half hour of aerobics Monday through Friday at home and weight training on Monday and Thursday at Headquarters, which gives me a chance to say howdy to my comrades and mingle with people. I clean the house and wash dishes as needed (I'm a good housewife) and take care of the yard and remove snow as required and perform home repairs. I know my wife really appreciates coming home to little or no housework after a full day at her job, which by the way seems to be a place just as busy and frustrating as the one I just left! Listening to her vent about her workday makes me realize that I wasn't alone in my day-to-day struggles and find my new life a stark bit different from my old life especially when listening to my spouse. I feel guilty because I have escaped that tier of experience and she has a ways to go before she is ready to leave her career behind.

The other day I was hiking in a place known as Butterfield Canyon in the southwest corner of Salt Lake Valley. I wound up hiking with a lady who had worked for and retired from the IRS about ten years ago; as we walked and got acquainted, she asked me if I ever feel guilty about being so free? I said I did; she said she did too even though we had worked most of our lives. I personally had worked for about forty-five years and had two jobs for most of that time and sometimes three. I told her not to be too hard on herself though because we both had earned the right to our benefit the hard way by working year after year one day at a time.

Lately, I have also been engaged with the doctors keeping my neck and back healthy. In December of 2013, I received a second injection warding off the full plate (again) that is hovering around like a ghoul. I engaged in writing this book on a regular basis soon after I retired and formed some ideas for two or three more books about the outdoors, which I would like to write soon after this one is done! I still field-test outdoor gear and have found that the younger generation of designers are interested in my input, so I feel that all important sense of being needed and will help those folks for a while. Testing will help get me up into the hills hiking and fishing or out on my road cycle enjoying the fresh air and the freedom a bit more too.

I get out with my telescope as the urge comes up and have been getting my grandkids involved with astronomy too. It's a great hobby and is quite fulfilling especially when you realize the scope of our life here on earth and then compare it to the incredible vastness of our solar system then our galaxy and then our universe. We have so much to learn about life beyond our planet—no question; this adventure will most likely require teams from all over the globe to accomplish very much in this arena! I also read quite a bit on all sorts of topics that I find interesting, ranging from music to science and adventure! So I guess I am becoming a writer and am more available to be a husband, a father, and a grandfather too! And as usual, I still have lots of people and variety in my life, which I think is one important key to surviving retirement!

I will say this: at the beginning of every week, I still wake up with a little panic, which I think is relative to the fear of boredom; maybe with time, this will go away, or is it a good thing—a reminder to keep active maybe? One doesn't necessarily have to race the wind all throughout life, but one ought to turn a rock or two over occasionally and look for new things to engage themselves in! Get away from the TV once in a while and go on a drive or a hike up a favorite trail, even on a winter day; it just might be refreshing. You should grab some binoculars sometime and look at the quarter moon, or swing them slowly through the Milky Way from a really dark site; you will see something truly majestic. Maybe you could call a friend you haven't talked to in a while and invite him/her along, just don't stagnate. I do get lonely and anxious, and believe it or not, this is one of the reasons I go to Headquarters to exercise two days a week; I need to hobnob with some of my friends, and it feels good. I don't feel any contention with this environment that was my final resting place in the workforce. I am treated well and am genuinely respected, and I feel proud that I was able to serve there and make a difference! The people there routinely comment on how good I look and want to know what retirement is like, so I share my thoughts gained thus far and encourage them to be prepared for it, to have the vitals in place, and to think about how they will spend their time after the fact! I look in the mirror and notice that I look better and feel younger and less worn-out. But as I look at my life, I can't believe how fast time has gone by and what I have lived through; then I realize that I am on the short end of the stick—no regrets though!

I reflect often about the past, hoping I have learned something that will help me do better in the future! I am so amazed with what was transpiring right next to me while I was stressing in my own little world, a very small space indeed, a space much smaller even than the circle my brother told me I could lord over! My wife for example had several serious health issues beginning in the mid-2000s. Her meniscus was damaged on both knees, which resulted in her needing a full knee replacement on her right knee! Teri had a great doctor and received the most technically advanced replacement knee at the time, yet she suffers from constant high-level pain! To add insult to injury, she grew a tumor above the new knee, which was affecting her feet and critical nerves in her right leg. Teri came home from work one day to change clothes; she was being interviewed by a local news channel about the school lunch program and wanted to look nice. As she was starting down the flight of stairs inside our home, her right leg quit working! She fell end over end crashing into our front door, fracturing several vertebrae in the process; I was scared to death for her and what this could mean for her future. She has not fully recovered from either of the two different events but keeps moving forward in life despite her physical limitations. We don't realize how much effort some people quietly put out just to get through what is a very friction-free day for others. Many people suffer from splits on the ends of their fingers, and she is one of those lucky folks who must bandage up their fingers 24-7. I know it's awful for her and feel bad for her because it hurts. We are forever trying to find a remedy—I keep my eyes and ears open for something that will help, but she keeps doing her job and being my wife, a mom, and a grandmother anyway!

Note: As of August 2014, Teri underwent surgery to rebuild her knee, and it has been, so far, a complete success. As it turns out, a sports medicine doctor (just taking a look) thought there was an issue with her kneecap. So her surgeon's intern took some X-rays, magnified them, and found at least some of the trouble (her existing cartilage had deteriorated). It's amazing; her pain level is mostly near zero, and she walks better than she has in five years—see, there is hope, you just have to believe and keep trying and listening. I suppose if I was going to learn patience and flexibility and discipline, I should learn it from her. I have always been in awe of this woman's toughness; she has endured so much just dealing with me. She has raised four kids, and she grandmothers over five children, helping them and loving them in every way possible. They are lucky to have someone who just simply finds the time and/or the pathway to take care of a need, all the while managing a lunch program for a major school district. She has an amazing ability to smooth out the wrinkles that life creates and is a great example of a selfless person and the term *giver*. I know this: our family is lucky to have her! Of course, there were so many other things going on in and out of my and her life, and I know she was lonely and scared quite a bit too—mostly because of me. I am so lucky that the bond she shares with me stayed strong through all the vacancies and other challenges that came along. We both agree that our faithfulness in these times made us stronger in our personal relationship with each other.

If it hadn't been for her careful planning for the last thirty years, my being able to retire may not have been possible at this point in time—or maybe at all. I have always been amazed at her intelligence, ingenuity, and her discipline with our budget! She finds the sales, is perfectly happy with her simple 2006 Toyota Corolla, she uses coupons, she gets the lowest interest rates, etc. All this helps stretch those precious dollars, which helps make many other things possible! So when I find out about something she really wants, she gets it, period! She is also a very good cook, is happy with the basic necessities of life, and she is a very beautiful woman who simply wants to be loved.

Other things I reflect on are how I have managed myself through most aspects of my life. Now that I have quite a bit of time to think things over, I realize that I should have taken more time to think things over! I think I could have been more flexible and patient with others; even though I think I was pretty fair with most folks, I should have been more fair, even if it didn't get returned. Ya know, when you look in the mirror, it is now all about you. I still, however, stand by some things as being constants in our ever-changing universe that I believe stand the test of time (i.e., everything that integrity defines)! I think it is wise to use your eyes and ears and your heart more than your yap! You and I both gotta realize just how important that statement and action is! Life is pretty out of control at times, and most people are really trying to do their best, even though there are exceptions? But ego is a very destructive thing and gets out of control so easy with us humans and can be super hurtful! People are posturing all the time because it makes them feel good and momentarily puts them on a personalized pedestal, which is balanced on the pea that resides inside their skull, which by size almost blocks the view through the ears. It is truly amazing how insignificant we really are compared to how significant we think we are; ego just glazes us over with self-importance until we fall off a six-foot ladder, breaking a shoulder and a leg, then we get pneumonia and live on a mechanical lung—then we die! Wow, how does this happen to someone so important and so significant! All we really are is just one of the many life-forms that exist on this planet, and if it weren't for our heads, we'd be okay; maybe we'd even outlast all other life, maybe we'd even get along. I realized quite a bit of this while traveling the world with the military helping folks in other countries—they are very much like us lucky-ass

Americans, just a little less fortunate given their particular set of circumstances! Like us, they want to be comfortable, they want families, and for them to be healthy, they even dream a little, they want peace and a decent bed to sleep in after their tummies get filled with some good food!

Let me tell you what, the things we take for granted is sickening, yet we are content with robbing even ourselves of the luxuries we fought for in the beginning and claim we defend right now. Looking back over my life, I wonder how I lasted so long and why I was so blessed or lucky (you judge). I think about all the different circumstances that I have passed through and survived—why, how! Many people have been killed dead by lightning just playing golf or standing by a tree. I had a far more severe set of circumstances, but here I am, alive—why, how come?

The military itself is replete with suicide victims, one was a personal friend of mine; why is he gone, and I am not? There were no signs, but now there is a void where this guy used to exist—I don't know… Recently, a cousin of mine was sucked into an asphalt grinder, a good kid, dead before he even knew it, and now there is another void—it's like the world keeps collecting these black holes where there once was light; the tangible just fades away, and a silent darkness takes over. It's like when I was dead (apparently), there was nothing; then, the light was given back to me. All I can say is that these particular states of being are completely opposite.

As I write this book, I am painfully aware of all the people dying around me and for countless reasons. People are also losing their mobility, losing their families and even their minds. Man, life is hard at times; it gives and it takes, and you can't stop it, just look around! All you can do is try to be a good person and recognize that there are many people right in your shoes and so many more far worse off! Just forget about the ones who seem to have it better—it'll drive you nuts with worry! The decade of the

2000s represented to me the hardest part of my life even without the lightning accident; so much was lost mostly in friends and family members and acquaintances. I think even part of my mind left too. But there were significant gains that came along also! Remember, life can largely be about perspective and our choices.

I was a different guy in the year 2000. Then I entered into a whirlwind where I ramped up my activity level and eventually lost perspective about what was happening all around me. It seems like I lost track of the big picture and of me and of who I was. I had a name, but my identity went MIA (missing in action). I often worried about the outcome of it all and where I was headed, but I kept going that way anyhow, why? Was I that lost or disturbed? What? I don't know—I used to know, I think…

A friend of mine during this time told me that God keeps you in a situation until the lesson is learned! Maybe, I told him, but I felt like socking God right in the eye over this one! But like I have iterated to death, I did learn some valuable things along the way despite the maylay. For example, it's amazing what we impose on one another; it literally can leave no room for God to exist. Many of these things we do are done out of fear of some kind; then there are things that are done out of pure meanness, which can be driven by all sorts of elements like greed, revenge, lust, substance abuse, and hatred—the very roots of evil. Unfortunately there is also a ton of misunderstanding in this world sitting right on top of it all too! Good criminy, help, please, Lord, help!

Luckily there are lots of things that are done out of pure love, of caring and understanding! Most folks are good, but we all have our faults too—yes, you and I do, no kidding—don't argue! Learn to be fair and consistent and don't judge too harshly! Hug your loved ones more often and tell them that you love them! Remember, life has its unpredictable side, and a lifetime is but a blink of an eye, just ask any dead person!

I need to say that overall, I am pretty happy with my life—even despite my personal set of negative experiences, things could have been far worse! As I see it, I lucked out in the long run. Just look at the family I have. I was fortunate to have had a loving mother and brothers and sisters, including all the steps and half siblings, my mother and father-in-law and their children too. My wife—she is my life! She and I have a great bunch of kids and grandkids, including the spouses; they take great care of us and are fun to be with. My wife and I are so proud of how they conduct themselves and their growing families! I also believe that my dad, my father, in his own way really does care a lot about me, and I care about him and have found peace to a degree with our history—it's okay if you disagree!

Reflecting on my history and with lots of time and water under the bridge, I've spent some real effort in understanding my pathway, the yin and yang of it, as Bruce Lee would say. So again, I am very happy with the way things have turned out and would not trade places with anyone as I have been very lucky indeed. I have had lots of second chances and opportunities to see and learn so many things, to adjust to life as it comes, and to avoid and/or learn about calamities big and small and to help other people. Yes, I have had a very good life! Could I have done better—of course, and I wish I could learn to communicate and to listen better. You know, it is never too late to change or right the wrong—it's simply a choice.

A few months ago, my youngest daughter, Amy, interviewed me for some medical people interested in how burn victims (like me) see things even years after their experience. I was startled at how deeply some of the questions dug and what I remembered! On several occasions I was submitted into remembering things that made me just sob. My emotional outpourings caused my daughter to ask if I wanted to end the interview. "No," I strained, tears dripping, "I just need a moment to get myself together." I felt that it was important for whoever is gonna use this interview to understand the depth of what I had experienced so they may help others get through their issues maybe a little more smoothly! This interview showed me how much I still hurt and how far I had come; it told me that I was still in need of help even years after the ordeal. Yet again and overall, life had been very good to me and mostly because of caring people, many of them didn't know me from Adam!

Many years ago while bouldering in the gate buttress area of Little Cottonwood Canyon, I noticed a familiar face trying to work a problem on a test piece boulder. Steve was one of the original "hard men" who had established many a route when the climbing scene here was first getting started. He had been gone from climbing for about a decade while attending college and had become pretty rusty and seemed frustrated with failing; over and over he exerted his will to make it up this boulder problem known as the Standard Overhang. While he was shaking off the previous attempt, I went over and talked with him for a few minutes for history sake then went on my way. Several months later, I learned that he had been killed in a skiing accident; needless to say, I was a bit shocked! His story stayed neatly tucked away in my brain for many years, resurfacing for edification on several occasions! I always wondered if he was trying so hard to relive the past that his emotions overrode his judgment!

Recently I had gone to REI (an outdoor store) to manhandle and look over a pack I was going to test for my buddy John at Gregory Mountain Products! While there, I visited with my old friend and expert climber and bouldering pioneer Ray! I hadn't seen Ray for a while and hadn't climbed with him for even longer (like thirteen years). We talked about all the great things we used to do in the mountains and how it felt to send away the tools that had made it all possible. We talked about our aging, injured bodies and other priorities of life that were there all along too! I told Ray that I had just sold my Quasar ice tools and Nova crampons; his eyebrows rose. Continuing, I said that even though I hadn't climbed seriously for nearly a decade, I was holding on to the dream kinda tight! I told him that as I drove the I-15 freeway through our valley, the canyons, Big and Little Cottonwood was in grand view, and it hurt a bit to remember the high places. Calmly my friend and mentor Ray said, "I know, I know." Apologetically he advised that the mind and spirit may be willing, but the mental and physical edge is gone; injuries these days show up out of the mere thought of pulling hard on a little crimper hold, that the body is just breaking down and steadily losing its capability, so we must move on. Then Ray asked what I had been doing since my retirement. "A little field-testing and writing a book," I said; he raised his eyebrows again asking what my book was about. I told him it was kind of a memoir and that it had a lot to do with the lightning strike and decision making and actions in a life's journey. "Hmm, good," he said. My friend Ray then told me about the state-of-the-art electronic equipment he had been working on in college and about his eighty-thousand-word thesis on the topic—this guy is smart! I reminisced about the climber I had talked with while bouldering who later died skiing, and how just the other day, I had woke up with him on my mind, now I understood something important about his situation relative to my own life! Ray wondered what it was that I had learned. "My friend," I vented, "I still have a great desire to be in the mountains with my ice axes," but I realized right then that I am not the hungry climber I once was and most likely neither was this other guy. I have been away from climbing a long time; I am rusty, and I don't like getting cold anymore and that I needed to move on despite the internal fire. I wondered if this desire I felt was just an emotion driven by past experiences. I knew if I went back, I would again have to solo, and I just might not survive it. I had had my time, and now with my new life (retirement), it was time to explore new possibilities, reevaluate my direction, get myself in order, and create some happiness. Ray agreed, hugged me, said "see ya," then went to help a customer!

I really don't know what it was that caught up with that old "hard man" climber, but from him in some way, I realized that I have to become wiser and follow my heart and that little inner voice that's been so good to me and not let my emotions steer me wrong either. I will try to make the most of the time that I have left and be as good as possible to my wife, my family, my friends, and all the other people around me! I guess it comes down to the things that I have the power to actually influence, and the question is, do I want that influence, like the ripples from the pebble thrown into the pond, to be positive or negative? Again, I think of that little circle that my brother taught me about. It's never too late to change direction, to right a wrong, or to give someone a hug or say you care.

A great man once suggested: Go into the woods and spend a night by a campfire alone, then focus on you and your life and all you have done, and figure out where you are headed. This is a great time to come face-to-face with some facts; you may even decide that it's time to institute some changes. Remember, life is a privilege and is shorter than you think, and when life ends, it is done, and what you've done or haven't done is in the books!

Good night.

Afterword

Who am I now, what did I learn, and is it too late?

I think about this a lot because I've had so many diverse experiences in my life, and I ain't done yet! The decade of my fifties seems to be the one where I was tested the hardest, and on many different fronts, it was also the last of my time in the military and civilian workforce. Now that I have been retired for several years, I have had some time to, as we say in the uniform, reconstitute—to become in balance with the home playing field again. I can see now that I could have been more open, even a bit more reasonable, even though I thought my flexibility grew as time went by, and I do think twice more often! I realized that there were things done that were wrong, and they will always tick me off because they will always be wrong, and yet they were allowed; this caused me to be hateful, but then there were all the right things that came along too. Mean people like the ones I worked for are often just misplaced individuals that may do better in a different environment, hopefully. One thing for sure is this: the test of time proves all things out—right or wrong—and you are left standing facing the reflection of your own acts. Nonetheless, the circle of influence says there are some things that will always be out of your hands, and that ticks me off too, but I can't argue with its correctness. I also think that sometimes having everything perfect can complicate life just as much as it can smooth things out.

I remember meeting a young soldier who was a patient at the hospital in Landstuhl, Germany, where our military unit was stationed during Iraqi Freedom. He had been on a recon mission just before the taking of Baghdad and was returning the way he went in and stepped on a land mine, severely damaging his legs. I noticed that while in his wheelchair, he was struggling to get a tube back into one of his destroyed legs, so I went over to assist him and thanked this brave twenty-two-year-old for serving. With a real concerned look, he thanked me and said that he respected me for serving also. I was humbled and told this young soldier that I have the greatest admiration for him and what he does for his career field or MOS (military occupational specialty); after all, he was a combat soldier and I am an electrician. This young man simply stated, "You went before me, sir, the respect goes to you first." I was in tears over the courage and discipline and respect this troop possessed. Meeting this young man reinforced my faith in the generations following mine—that there is still a code of ethics in our country and that there are still a people willing to stand in the gap for what is right!

Thinking out loud now, to me, finding closure and becoming grounded can be about learning to forgive and sharing yourself and always doing your best—no matter what, no excuses. Treasure your family and your time on this earth, not your possessions. The struggles that held me back eventually moved me forward and caused me to search for me and who I am.

I guess I still have a few more things to think about and to reconcile with and adjust.

Lastly, I need to relax a little.

Is it too late to change? It's never too late.

About the Author

The author served for thirty-two years as an electrician for the Utah Air National Guard, deploying all over the world in the uniform and served as a civilian electrician and safety trainer for the Air and Army National Guard for nearly twenty-four years.

He has spent a great deal of time searching and learning about the night sky with his telescopes and getting other people involved in this great adventure. He has climbed many technical rock and ice routes and has backpacked and hiked over much of the Western United States and has spent two decades testing outdoor-related equipment for dozens of manufacturers, including teaching lightning-awareness classes to a variety of people, including the Boy Scouts.

Presently he lives in Utah with his wife of thirty-four years; together they have four children and five grandchildren.

Most likely you will find LT, as he is known, out hiking on any one of his favorite trails in the Wasatch Range near his home.